# Fundamentals of Aviation Operations



This book provides a general introduction into aviation operations, covering all the relevant elements of this field and the interrelations between them.

Numerous books have been written about aviation, but most are written by and for specialists, and assume a profound understanding of the fundamentals. This textbook provides the basics for understanding these fundamentals. It explains how the commercial aviation sector is structured and how technological, economic and political forces define its development and the prosperity of its players. Aviation operations have become an important field of expertise. Airlines, airports and aviation suppliers, the players in aviation, need expertise on how aircraft can be profitably exploited by connecting airports with the aim of adding value to society. This book covers all relevant aspects of aviation operations, including contemporary challenges, like capacity constraints and sustainability.

This textbook delivers a fundamental understanding of the commercial aviation sector at a level ideal for first-year university students and can be a tool for lecturers in developing an aviation operations curriculum. It may also be of interest to people already employed within aviation, often specialists, seeking an accurate overview of all relevant fields of operations.

**Gert Meijer** is a lecturer in Aviation Operations at the University of Applied Sciences in Amsterdam. He started his career in aviation in the sales department of aircraft manufacturer Fokker and has worked for Honeywell Aerospace, where he was responsible for sales and support in commercial aviation. He holds a Master's degree in Political Sciences from the University of Amsterdam.

## Aviation Fundamentals

*Series Editor: Suzanne K. Kearns*

Aviation Fundamentals is a series of air transport textbooks that incorporate instructional design principles to present content in a manner that is engaging to the learner, at an accessible level for young adults, allowing for practical application of the content to real-world problems via cases, reflection questions and examples. Each textbook will be supported by a companion website of supplementary materials and a test bank.

The series is designed to help facilitate the recruitment and education of the next generation of aviation professionals (NGAP), a task which has been named a 'Global Priority' by the ICAO Assembly. It will also support education for new air transport sectors that are expected to rapidly evolve in future years, such as commercial space and the civil use of remotely piloted aircraft. The objective of *Aviation Fundamentals* is to become the leading source of textbooks for the variety of subject areas that make up aviation college/university degree programmes, evolving in parallel with these curricula.

**Fundamentals of International Aviation**
*Suzanne K. Kearns*

**Fundamentals of International Aviation Law and Policy**
*Benjamyn I. Scott and Andrea Trimarchi*

**Fundamentals of Aviation Operations**
*Gert Meijer*

For more information about this series, please visit: www.routledge.com/Aviation-Fundamentals/book-series/AVFUND

# Fundamentals of Aviation Operations

Gert Meijer

Routledge
Taylor & Francis Group
LONDON AND NEW YORK

First published 2021
by Routledge
2 Park Square, Milton Park, Abingdon, Oxon OX14 4RN

and by Routledge
52 Vanderbilt Avenue, New York, NY 10017

*Routledge is an imprint of the Taylor & Francis Group, an informa business*



*British Library Cataloguing-in-Publication Data*
A catalogue record for this book is available from the British Library

*Library of Congress Cataloging-in-Publication Data*
Names: Meijer, Gert, author.
Title: Fundamentals of airline operations / Gert Meijer.
Description: First Edition. | New York: Routledge, 2020. |
Series: Aviation fundamentals | Includes bibliographical references and index.
Subjects: LCSH: Airlines—Management. | Business logistics. | Scheduling. |
Aeronautics, Commercial—Passenger traffic.
Classification: LCC HE9780 .M45 2020 (print) | LCC HE9780 (ebook) |
DDC 387.7068—dc23
LC record available at https://lccn.loc.gov/2020012459
LC ebook record available at https://lccn.loc.gov/2020012460

ISBN: 978-0-367-33240-2 (hbk)
ISBN: 978-0-367-33239-6 (pbk)
ISBN: 978-0-429-31880-1 (ebk)

Typeset in Minion Pro
by codeMantra

# Contents

# Figures

# Preface

This textbook is about aviation operations: *the organizations and infrastructure in which commercial aircraft operate between airports, providing connections to society with the aim to earn profits.* It is about rules, markets, suppliers, airports, airlines, money, people, politics and the aircraft itself. Aviation operations is a sector in which commercial aircraft are operated, deployed, assigned to routes, planned and scheduled from and to airports that are themselves tightly planned and scheduled. And all with the purpose of *adding value to society* as a transport sector.

In the old days, aviation was formed by engineers, pilots and traffic controllers as the aviation professionals, and thus aviation organizations were run by staff originating in either of these three professions. In the old days, airlines and airports were run by pilots and engineers. Occasionally economists and lawyers got involved as outsiders, but lacking aviation knowledge they were hardly in charge. The CEOs were still engineers, pilots or even astronauts.

With aviation becoming a *mature profit-driven service industry*, this is changing dramatically. Aviation operations have become a field of *expertise unto themselves*, based on *economics, statistics and logistics*. Aviation operations involve the technology of aircraft, rules and regulations, business strategies and political ideologies. Aviation organizations today have an increasing need for operations staff that is *educated for operations*. Operations expertise is needed to cope with contemporary complexities, with the need for profitability and with increasing capacity constraints.

This textbook is meant as a first introduction to all these relevant aspects of today's aviation operations. It is primarily meant for the *student* with high-level ambitions in aviation, resulting from a keen interest in aviation and a robust high-school level but zero or limited *knowledge* about the complexities of aviation. As such this book aims to form a basis for further bachelor's or master's level studies in aviation operations. This implies that, although meant as an introduction of the fundamentals, the content of this book is *not* simple and not limited to a rudimentary level. After digesting this content, the student has a broad and solid basis for further reading on selected topics. This book aims to introduce all truly relevant topics as well as the legalities, the technicalities, the business and the operational challenges of aviation operations.

This book is also written for *lectures* in aviation operations. *Being one myself*, I found it difficult to find suitable educational operations material for students new to aviation. There are many books published on aviation operations, but most, if not all, are written at a sophisticated level, *for professions and by professionals*. My first-year students encounter difficulties in accessing and digesting all that high-level knowledge, leaving me, as the lecturer, with the task of developing material to bridge the gap between the starting level of the student and the high level of the literature. *This book aims to provide such a bridge*. Its structure is meant to enable lecturers to straightforwardly compose lecture material. This makes it easier for educational institutes on aviation to develop operations as a field of expertise next to the technical curriculum. And even in a technical curriculum this book may fit. Employers in aviation prefer to recruit flight operational and engineering staff that understand aviation as a sector as well as how profits are made. Employers in aviation operation need new staff educated in the expertise of aviation operations.

This textbook may also be of interest to aviation *professionals*. Odd as that may sound, it is my experience in over thirty years working in aviation that the field is run by *specialists*: highly qualified specialists with a profound understanding of their own specialism but sometimes limited or coloured perceptions of other specialists in aviation. This book may help specialists to better understand other specialists and their colleges in other departments or organizations. After all, the more various specialists understand each other, the more effective aviation can perform. And in today's society, aviation operations need to perform to the max.

The author of this book is anything but a specialist. From early childhood, I have had an indefinite passion for aviation, and as a Political Scientist I am intellectually trained to understand integrated constellations formed by technology, economics and law, all glued together by the notion of human nature, culture and time – like states or governmental organizations. I have translated this academic thinking to the field of aviation. The combination of political scientist with a certain technological understanding of aircraft has been fundamental to my aviation career. I have worked in marketing and sales on the supplier side of aviation – responsible in various companies and on various continents for the sales and support of aircraft, engines and their components. It taught me not only the technical specifics of aviation but also how the airlines use these products and how they make their profits for survival – and how they all do that *uniformly*, on the basis of tight international regulations, versus how they all do it *differently*, based on traditions, cultures and believes. Working for and with aviation companies worldwide made me a *generalist* in aviation. This textbook is a product of this generalist view; modern language might call it a *holistic* view.

This holistic view includes the *business side* of aviation. Often disregarded by standard textbooks on aviation, and limited in perception and attention by the academic world, I have learned that commercial aviation is foremost a business, and one needs a keen understanding of the dynamics of the multi-billion-dollar business that aviation is in order to understand the dynamics of aviation operations.

The holistic view includes the *legal basis* for aviation as well. All parts of this book will start with the legalities of the subject of that part as understanding the legal framework is indispensable to understanding aviation operations. One needs to know the rules to understand the game. But one also needs to understand that all this rule-making is far from value-free as the rules and regulations governing aviation reflect the political ideologies of the authorities that make these rules.

This textbook covers the field of aviation operations, often regarded as opposite to aviation engineering. Engineering is about how aircraft are designed and constructed, how they fly and how they are maintained. Operations is about how aircraft are deployed by airlines to airports and in the air under the guidance of ATC (*Air Traffic Control*). It is about how technology can be applied against cost affordable to many. This implies that operations and engineering hold a close relationship in aviation, that they are two sides of the same coin. Understanding one side requires at least a basic understanding of the other side. This textbook therefore covers the relevant technological elements of aircraft as well, wherever necessary for understanding how aircraft are deployed in aviation operations.

Writing this book was only possible with the help of many. The management team of the Aviation Academy enabled me to write this book. I need to thank John Wensveen, Paul Clark, Bob Walton and Geert Boosten for their critical comments on my text. I am grateful for the help provided by Raymond Teunissen as he corrected or improved my draft texts on the technical aspects covered here. Finally, I thank my former colleague Pieter van Langen; he provided the moral pressure that was indispensable to actually finishing the book.

Amsterdam, February 2020

# Introduction

This textbook is about commercial aviation operations, about how aircraft operate between airports transporting passengers and cargo. Aviation is an indispensable transport sector in today's *globalized* society; it is, like the internet, an *enabler* of that globalization. Aviation enables the physical transportation of people and goods over long distances within short time frames. It does this fairly cost-effective, compared to surface transport modes. In today's society, it is for many people possible to spend a week's holiday on a tropical island, to visit a customer in another continent or to buy fruits in the supermarket that was harvested in other parts of the world only days ago. In supply chains, aviation takes care of the distances, enabling global trade. Aviation is, next to other technology sectors, an enabler of the global village, the global socio-economic interrelations between modern societies that characterizes the era that we live in.

The vast geographic focus in our everyday lives is very specific for our times. In the past, distances were difficult to concur and the world was organized in local societies that were mainly isolated from each other. It took mankind a long time before the motorized and controllable aircraft was invented by the Wright brothers in 1903.

The tremendous advantages of this invention were soon recognized. An aircraft flies at high speeds in a linear direction, and *without the need for infrastructure* on its route. This enables transportation to far-away destinations in short time frames. Aviation replaced maritime vessels for passenger transportation business on the advantage of speed. And as the earth is covered by water, deserts and mountains, the autarkic operation of aircraft in the air soon became the basis for a wide proliferation of aviation around the world.

And still today, the speed and independency for ground infrastructure make aviation a transport sector that truly adds value to today's world society. As a mature industry, air transportation has become a commodity, and that means production and consumption at high volumes. And with these volumes came the setbacks for aviation. Its need for fossil energy, its environmental impact, and the capacity constraints of its infrastructure: airports and air traffic control.

In order to understand the development of aviation to how aviation is structured today, we will briefly look back at some 100 years of aviation development, touching upon the main dynamics for change and development. This book is not meant as a history book, and bookshelves can be filled by books on the history of aviation. However, some steps in history are important for understanding today's aviation sector, and for understanding the fundamentals for the development of aviation. Let us start with a bit of history, understanding the time-line of aviation.

Aviation started with the emergence of the *combustion engine*. This step enabled the Wright brothers in 1903 to get their Wright Flyer in the air and control it. Young men around the world were sparked by this and built their own engine-powered controllable airplanes. Many died in the action; few succeeded. The pioneers were born. Their creations were machines built of wood and linen, and later, aluminium tube fuselage frames. Their aircraft were capable of uplifting itself and the weight of the pilot and some gasoline in a small tank. Usage for these machines was quickly found in warfare. Thousands of aircraft were built and destroyed in the *First World War*, the "delivery room" of aviation. The aircraft was born.

The colonial powers in Europe were the next to discover the breathtaking potentials of aviation. *Contact with the colonies* could be reduced impressively by transporting mail by airplane, quickly followed by the need for transportation of persons as well. By using engines optimized for aircraft and applying a hollow plywood wing that could carry significant masses and store gasoline, the first pioneers constructed aircraft that could carry up to ten passengers over distances of nearly 1,500 km. With these machines, newly formed national flag carriers in Europe started in the early 1920s to fly between the mother country and their colonies. (These routes are still the largest airline markets in today's industry.) Rules and regulations appeared first on national levels, but quickly, international rule-making emerged by means of multilateral treaties like the Paris convention. Capital cities started constructing airfields. The *civil aviation sector* was born.

In the early 1930s in the US, the air-cooled radial engine was engineered. Applied in an airframe made out of aluminium, US-based pioneers started constructing aircraft like the DC-3, which could carry up to 30 passengers some 3,000 km, comfortably and relatively safely. These new qualities fuelled the development of the US domestic aviation market (still the largest single airline market in the world). The aviation sector became a *commercial business*.

Then came the Second World War, with a need for long-range bombers flying at high altitudes. The pioneers engineered them, by applying turbochargers in the engines allowing combustion engines to operate at higher altitudes. At these altitudes aircraft fly faster while consuming less fuel. The pioneers built them in massive numbers, capitalizing their start-up enterprises into firm companies like Boeing, Douglas and Lockheed in the US, and Vickers and De Havilland in Britain. With their post-war models based on the emerging technologies, they founded the post-war development of long-haul aviation exploitation. *The long-range transport* was born.

Aviation became a significant sector in the world, as from 1944 internationally regulated by the UNO agency for civil aviation, ICAO. Airlines, national flag carriers, were a proud appearance of the nation state. Internationally, airlines connected their capitals; in larger countries a domestic system appeared. Aviation became a competitor for trains and ships. Aviation provided fast and safe transport, be it against high cost. Its clientele consisted of the higher classes, the rich and the famous. In this period aviation got its *glamourous reputation*.

**Figure I.1** Douglas DC-3

Airports became the highly visible gateways for cities, often with high ranking architecture. Airlines became the main international appearance of their nation. Aviation has become a very visible and important sector, a sexy industry.

Behind the drawing tables in the meantime, it was concluded that the piston engine was at its end, and the turbojet- and the turboprop engine were engineered, based on knowledge and experience from WW2. Generating enormous amounts of power, turbojet engines enabled the emergence of aircraft in the late 1950s that could carry some 150 passengers over distances of up to 6,000 km at a speed of some 900 km. These aircraft, the Boeing 707 or the Douglas DC-8, got wings swept backwards. Flying times between continents were halved and could often be performed non-stop. Productivity went up, and aviation won the battle for long-haul passenger transport. With the disappearance of the passenger ships, the world became *dependent on aviation* for intercontinental connectivity.

The turboprop engine found its application in short/medium haul aircraft, enabling airlines to develop domestic, regional and continental networks. This dual application of the jet principle, the turbojet for long distances and the turboprop for shorter flights paved the way to airline networks with the capital as home base in the middle. Airline *networks* got structured. The turbojets for long distances were almost solely produced by US companies, with European-based suppliers holding secondary positions in the market for short-haul aircraft.

The turbojet had an inherent efficiency-reducing problem: *hot exhaust air colliding with the cold outside air*, causing jet exhaust turbulence at low altitudes, especially in take-off, eroding take-off power. Adding a big fan at the front of the engine whereby high amounts of air are led around the hot engine was the solution. This bypass air is warmed up and forms a smooth border between the exhaust air and the outside air. The *high bypass engine*, doubling the usable amount of engine power, was engineered.

This led to doubling the capacity of the aircraft. With up to 350 seats the Boeing 747, accompanied by other wide-body models, took over from their turbojet powered predecessors in the early 1970s. Flying cost *per seat* dropped drastically, and so did the price of tickets. Aviation became a mass transport sector. Aviation now created its own demand, basis for quick and solid growth.

In this period, newly independent nations started to develop, and with the emergence of Asian airlines the airline business became global. Airlines from Europe, the US and Japan now had to compete against airlines originating in countries with lower costs, mainly for labour. Furthermore, these new entrants introduced modern management and marketing techniques that the technology-driven flag carriers had not adopted so far. By applying these techniques, the Asian carriers introduced operational efficiencies, and by focussing on customers they developed new service standards. These new entrants, think of Singapore Airlines or Thai airways, became fierce competitors. The airline and airport business has become *global*.

As the large long-haul aircraft were actually far too big for the home demand of many national flag carriers, many of these carriers started to operate a *hub-and-spoke network* system, connecting their flights on the home airport, for feeding the too large long-haul aircraft. This has led to many airlines being bigger than their home market would allow. Their hub home airports became large and important points in international aviation connectivity. The *hub airport* was born.

**Figure I.2** Boeing 747

**Figure I.3** A380

The high-bypass technology has evolutionary developed since then; engines became more powerful and found their way to smaller, narrow-body, short-haul aircraft.

By reducing the number of engines from 4 to 2, thereby reducing the aircraft's weight, and by reducing fuel burn in order to allow for higher payloads, the cost per seat on today's long-haul wide bodies has *halved* since the introduction of the Boeing 747. And by applying four of those engines to other aircraft, a mega jet like the Airbus A380 has become possible.

Emerging technologies and the widespread use of digital technologies have decreased the labour factor, both in quantity but also very much in quality, to the extent that one truly can earn money by exploiting aircraft. Supply of highly efficient aircraft enabling this is done about equally between the US and Europe.

The commercial attractiveness of aviation has resulted in the creation of a multitude of airlines and airports. Even newer entrants from China and the Middle East, all operating from high capacity hub home airports, have entered the long-haul airline arena. On continental markets, the low-cost carriers have emerged, flying not only between capitals but predominantly between regional points. In today's world, we can fly from anywhere to anywhere against a fare that a vast percentage of world population can afford.

Looking back, we can conclude that regarding the technological dimension, the pace for the development of aviation has always been set by *engine technology*. Fuel-efficient engine power enabled the creation of efficient aircraft.

The economic dimension is that cost effective technology enables *profitable exploitation*. As margins have remained low, *success is found in quantity*. This quantity is fuelled by vast economic developments. Aviation has grown big as other sectors like automotive or IT.

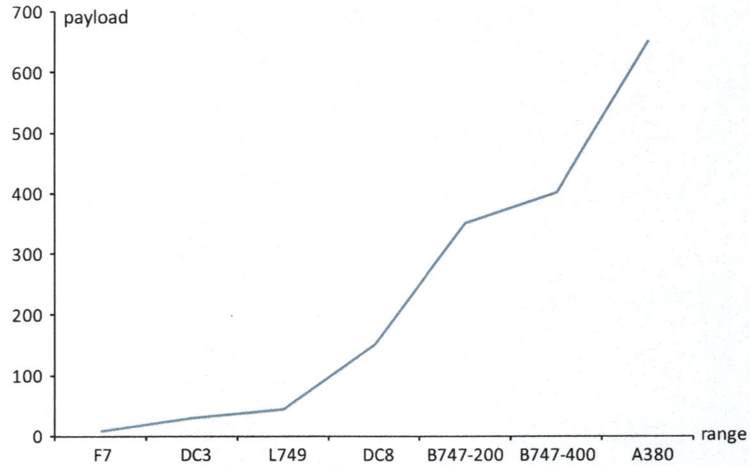


**Figure I.4** Productivity thru engine development

However, contrary to many other sectors, aviation has become an economic factor not only by serving the global economy but also by enabling it. Aviation has been able to *create its own markets.*

On the political dimension, we can conclude that the impressive development of aviation has been enabled by a balanced difference between robust *rules and regulations* for safety and security, and *liberalization or deregulation* for the commercial exploiters of aviation.

What we learn from looking at the history of aviation is that aviation is based on technology, economy and politics. This textbook will try to determine how this has all worked out in aviation as it is today, and how it has formed aviation operations as a relevant field of expertise.

# PART I

# The aviation sector

This textbook aims to introduce readers to the fundamentals of aviation operations. But before we can turn to the operations of aviation, we need a basic understanding of what aviation is and how this sector is structured. We therefore start with introducing the structure of aviation, and explain how this sector is organized. We will see who the different parties are in aviation, and what the technological, legal and economic fundamentals of this sector are.

One can look at aviation from very different angles. Technicians may primarily see the technological aspects, whereas economists see the vast amounts of money involved. For communities, aviation is a sector that connects their community with the rest of the world. Consumers see aviation as a service industry providing transport to far-away destinations. Aviation is indeed all of this; it is a sector based on *many fundamentals*. It comprises a multitude of complex *technologies*; enormous amounts of *capital*; and, because of its need for *safety* and its *international* character, vast *political* involvement. And, above all, it is all done by *humans*. People design, produce and operate aviation, whereby they all need to trust each other, in different roles and within different organizations. Many of these organizations are led by *market demand* and the need for *cost effectiveness* as well as the need for *safety*. And as a growth sector limited by many factors, all these different organizations need to act in concert for optimizing aviation operations both in *quantity*, by means of optimal deployment of capacity, and in *quality*, by producing reliable and robust air transportation based on demand.

In Part I we will look at the aviation sector from various angles in order to understand the various entities in aviation, all focussing on their specific tasks. We will first look at the three dimensions of aviation – technology, economy and politics – and see how they form the parties in aviation.

All fundamentals of aviation are based on either of these three dimensions, and often on a combination. When in aviation a player, be it an airline or an airport, intends to do something new, their first three questions are always: is it technically feasible (technology), is it allowed (politics) and does it make sense (economics)?

# CHAPTER 1
# Theoretical framework

## 1.1 Many different parties

The aviation sector consists of many profoundly different parties, all with their specific tasks and characteristics. Parties providing products or services to the passenger are very visible and known, like aircraft manufacturers, airlines and airports. Behind the scene there are many more, like Air Traffic Management, ground handlers or leasing companies, and in this textbook, we will take a look at these entities providing indispensable services behind the scenes. Airlines and airports are covered in Parts II and IV of this book.

Crucial for understanding how aviation functions, and for success of aviation, is the level of *cooperation* between these parties. Obvious as it may sound, it is far from easy since all the different parties have different interests and different priorities. But cooperation is a necessity; aviation cannot operate when the different vital parties do not act in concert. *Understanding* each other's differences is the basis for constructive cooperation in aviation.

One of the fundamental differences among parties in aviation is between *governmental* bodies versus *commercial* entities. Governmental bodies concentrate on safety and security without a responsibility for profitability. Commercial entities need to operate profitably, with bankruptcy as the final consequence for not meeting this basic requirement. The visible entities, like airlines and airports, provide services to the passengers and are therefore service industries in consumer markets; the entities behind the scenes have a different scope.

A closer look reveals that the different parties in aviation differ also in their development and dynamics. Governmental bodies are guided by politics, airlines and airports are guided by economics, and all are guided by the constant development of technology. If we look at aviation from this perspective, we see the three dimensions that its structure is based upon: technology, economics and politics. These dimensions are connected by the human factors.

### 1.1.1 The technological dimension

Aircraft are very peculiar machines; it took mankind centuries to invent them. An aircraft conquers gravity by deploying large amounts of energy. It needs to keep moving to fly and cannot stop in an emergency.

It needs to be strong enough to fly at high speeds and high altitudes. It needs to protect its occupants by generating and maintaining a liveable environment while flying in atmospheric conditions that are lethal to humans. And, above all, it needs to be as light as possible in its constant energetic fights against gravity.

An aircraft is a machine involving physics, mechanics, metallurgy, chemistry and electronics. In order to understand the technology of an aircraft, we have to divide this into three parts: *airframe*, *engines* and *components*.

The airframe is based on applied physics. *Aerodynamics* define the aircraft's ability to fly. *Mechanics* determine its strength, whereas *metallurgy* and increasingly *chemistry* define its construction. *Electronics* defines its operation.

The engines are based on the same combination of these technologies. Here, thermodynamics defines the airflow, the basis of jet engine power. Coping with the high temperatures inside the engine requires metallurgy and chemistry. When coping with the high forces mechanics is crucial. Combustion is defined by thermodynamics and chemistry and it is all controlled by electronics.

The components, all devices that are installed somewhere in or at the airframe, can themselves be classified as mechanic, hydraulic, pneumatic and now increasingly electric, all controlled by electronics. Avionics is predominantly defined by electronics in our digital era.

An effective aircraft needs these technologies in well-balanced combinations, whereby every aircraft design is a compromise between optimal application of each of the different technologies involved.

These complex machines need to be maintained, requiring technical skills to work with all these different technologies.

The essence of flying is to conquer gravity at a high velocity. If it fails it will fall to the ground, fast and hard. This makes flying inherently dangerous, reason why *safety* and *human factors* are crucial sciences for aviation.

Aircraft design requires experts on each of these technologies. Designing an aircraft requires many years of research and development and extensive testing. This design is finally produced by sophisticated materials and production techniques. All this combines to make aircraft *expensive* machines.

### 1.1.2 The economic dimension

Aircraft are indeed expensive machines to design and produce, and also to operate. Aviation therefore is a *capital-intensive* sector. Operating aircraft, in the air but most notably on the ground, at airports and hangars, requires many people. Aviation is therefore also very *labour intensive*.

Aviation is a truly global sector, where the money streams start with the customer paying for transport. In a liberalizing global economy, this makes aviation a very *competition intensive* sector, requiring *optimal cost effectivity by all players* and at all parts of the value chain.

Designing, producing, owning and operating aircraft require *financial economics* as the aviation business is a multi-billion-dollar business. Exploiting aircraft and airports require profound knowledge of *business administration* and *micro economics*, and *transport economics*. On the supply side, managing *cost* is the economic core.

Understanding and exploiting demand for aviation require *customer marketing & sales* knowledge. On the demand side, managing *income* is the economic core.

Keeping the income higher than the cost is the essence for economic sustainability. In aviation, this is a true challenge, and the economic dimension of aviation is therefore characterized by *low margins*.

As aviation is a global transport sector, it reacts fast and fierce to m*acro-economic* developments. The aviation sector therefore is very *cyclic*, with steep upturns and deep crises over time.

Economic entities and actors, like countries, regions or cities but also enterprises and organizations, need to be connected. This can be by phone; by data link; or by physical transport modes, like cars, trains and aircraft. Aviation therefore is of *crucial economic importance* for countries, regions or cities.

### 1.1.3 The political dimension

The above teaches us that aviation is technology-, capital- and labour-intensive. Aviation is also a sector with huge impact on regions and societies, a source for high educated employment and a creator of vital connections. Flying is inherently dangerous with fatal consequences if things go wrong. Here we have the ingredients for the *vast* political dimension of aviation.

Governments are a prime *finance source* of aviation as owner or subsidizer of aircraft production and airline or airport operation. This governmental involvement is considered necessary as economic margins are low and financial risks are high.

In the first decades of aviation *direct involvement* was the normal modus. With strong political forces towards liberalization and deregulation at many regions in the world as from the end of last century, the *indirect forms of state financing* have become more important. Indirect financing in aircraft design and production takes place via direct financing of defence or space projects with *technical spin-off* to commercial projects, state *guarantees* on commercial loans, *scientific* R&D budgets, *tax relief* on design and production cost, and *finance support* on sales transactions.

The political motivator for this financing is *industrial politics*, with labour market, technological capability and export market considerations for the nation acting as key drivers.

For the so-called super states, technology and most notably *defence* aerospace is key to their status in the world. Super states like the US, China, Russia, India, France and Britain finance their aerospace sector for political reasons. Maintaining their status as super power is the key driver here.

Airline and airport operators can be *state owned*, still common in countries where state involvement in society is high. Cities or regions also can own airports. In deregulated regions, airlines often are supported by discounts on taxes or charges. Even in liberalized countries governments often are *shareholder* in their national airline(s).

In operations, governments are the enablers; they sign *aviation treaties* among each other that allow airlines to conduct flights to and from their territory, or to overfly it, making use of its airspace. As such, governments *regulate aviation operations*. They can stimulate the sector, or protect their national entities against competition.

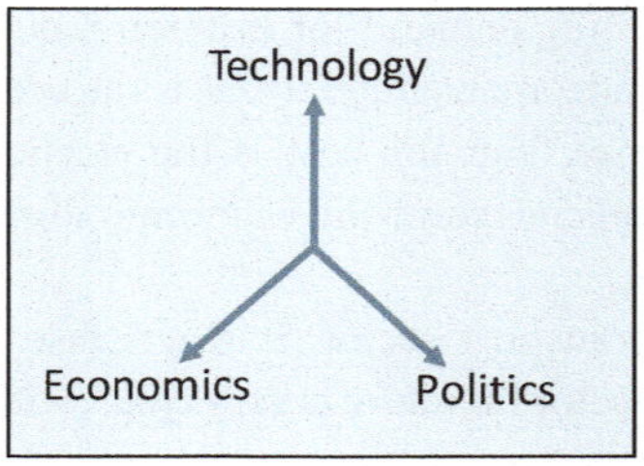


**Figure 1.1** The three dimensions of aviation

In the air, as well as on the airports within its territory, the state is the provider of *Air Traffic Services*, a crucial element in the total aviation value chain controlled by governments.

As aviation is a potentially hazardous sector, governments impose tight *rules and regulations on the safety of aviation*. Governments formulate these rules and regulations and enforce them, with the power of harsh penalties when violating them. Aircraft, airports and all staff directly involved in the operation are subject to a wide range of rules and regulations, ensuring an optimal safety of aviation.

Since the early 1970s, aviation has become the victim of terrorist activities, which is why governments impose detailed *rules and regulations on the security* of aviation. Based on the sound principle that an aircraft need to be secure before starting its flight, this security enforcement predominantly takes place at airports. At airports with international connections, governments are also present, executing their tasks via *border control* and *customs*.

As aircraft are profound producers of noise and emissions, both aircraft and airports are beholden to often tight *rules and regulations on the environmental impact* of their operation.

## 1.2 The human factor

Aviation is made by humans. They design aircraft and airports, they operate them, guide and control them, all within organizations run by humans. Aviation is by definition team-work, where people from all over the world need to work together. This requires common language and common rules, but above all, it requires *trust*. Aviation can only operate by professionals trusting each other.

Mutual trust is a binding factor in aviation at all levels. Pilots need to trust the mechanics, and airlines need to trust their suppliers. Even states need to trust each other; an aircraft from one nation should be able to fly and land safely in the territory of another state.

Humans operate the complex technology. They need to understand this technology and need to act and react in accordance with what the designers of that technology had in mind. The human pillar of aviation therefore also involves the man-machine interface, basis for the important subject of human factors in aviation design, production, maintenance and operation.

## 1.3 Dynamics of aviation

The aviation sector is constantly changing; the influence of the three dimensions that we described makes aviation a very *dynamic* sector. The three dimensions of aviation – technology, economy and politics – influence each other constantly. This results in *internal change factors*.

But aviation is part of the world, so changes in the world influence aviation as well. *External change factors* can also have a big impact.

Changes in *aviation technologies* is the main internal factor for change, often changing the economic parameters of aviation operations. The introduction of a new aircraft type can change the network of airlines in the destinations and frequencies they fly. The introduction of the Boeing 747 in 1969 lowered the cost of flying per seat, boosting the growth of long-range aviation markets. Airports constantly adapt to these changes as well, by adapting their infrastructure to receiving larger aircraft.

Rules and regulations change regularly, often as the outcome of incidents or accidents, or as a result of newly applied technologies. Aviation is a sector of constant change.

The main external factor for change is *economic growth*, creating new aviation markets, and economic changes or shifts in the economy changing aviation markets with it. The explosive economic growth of China has created new aviation markets, both domestic and international. The massive (labour) *migration* – domestic, regional and global – creates vast aviation markets in their slipstream. *Urbanization* creates new destinations for airlines.

Aviation is also very adaptive to *external technological developments* like the development of digitization, the emergence of the internet or the application of new materials. Due to the large amounts of money involved, aviation was an *early adaptor* of these technological shifts as initial high investments are recovered quickly in this multi-billion-dollar business. Efficiency improvements, weight decreases or cost reductions resulting from external technological developments are quickly applied in a competitive sector like aviation.

Contrary to all of the above, the aviation sector is often remarkably *traditional,* due to its addiction to *known* technologies, *proven* procedures and *familiar* business models.

The large amounts of money involved, the predominant need for safety and the long lead times for developing new technologies make the parties in aviation cautious. New technologies and new business models are not introduced before thoroughly proven. Often, these new technologies were applied in other sectors first or, when this new technology is aerospace specific, it has been applied in defence or space projects. Only when proven *reliable, robust and cost effective*, new technologies, procedures or business models are applied in aviation.

Also, the *long life-cycle* of aircraft, airport constructions and ATC infrastructure have a huge influence on the pace of change. After some five years of development, an aircraft may be produced and if so, it will remain flying for some 25 years, be it with modifications during this period. Airport construction takes years and become artefacts for decades. Upgraded and extended over time they will mostly remain at the location where its initial construction started, as a patchwork of different architectural periods.

Therefore, if we look at the innovative force of aviation, we always have to consider this careful *balance* between the advantages of innovation versus the need for proven reliability, within a long cycle life of aviation's major cost drivers.

Aviation also changes because of *geo political factors*, like the price of fuel or the political status of regions. As aviation is a major consumer of fossil fuels, the volatility of fuel prices influence the actual growth rate of the sector. After the demise of the Soviet Union, its

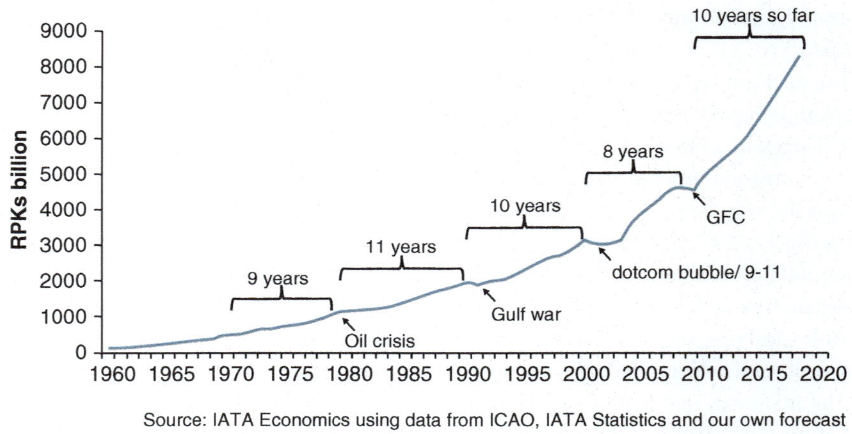


**Figure 1.2** Aviation reaction on external factors

massive northern airspace became available for commercial aviation, reducing flying times between Europe and Asia by many hours. This entirely changed the position of an airport, Anchorage, as well.

Thus, with strong internal- and external-change factors, aviation is a truly dynamic sector, in which all players need to change constantly.

## 1.4 Managing and planning capacities

In Section 1.2 it was mentioned that aviation is a business operating at low margins. Equipment and infrastructure require high investments to be paid back in time by transporting passengers and cargo. Aircraft need to be full for profitability; an empty seat is a waste. Airports need a steady throughput of passengers and cargo; an empty terminal does not pay back the vast investment needed to build it.

Available capacity needs to be deployed at maximum utilization to be cost efficient. For this reason, aviation operations need sophisticated capacity planning. *Capacity management* – detailed planning of capacity deployment – is a prerequisite for a successful operation of airspace, aircraft and airports. It is fair to say that next to the technological complexity of aircraft, the complexity of capacity management is the basis for aviation operations being a high sophisticated sector, requiring highly educated people to run it.

The challenge for capacity planning in aviation is that capacity needs to be deployed in accordance with market demand for transportation, and this demand is not equally spread. Not in geography, not in time.

Transport demand between mega cities like New York and London is very different than demand between regional communities. In highly developed nations like the US or Japan, demand is a much larger than demand in underdeveloped countries. These vast geographical differences are the basis for *strategic capacity planning*: long-term planning of capacity to be deployed tomorrow.

Within geographically defined markets, demand fluctuates in time: during the year, during the week and during the day. In the summer season or at Christmas time, demand is higher than in the early months of the year. Weekly demand peaks are on Mondays and Fridays, and demand peaks on a daily basis are in the early morning and the early evening.

On top of this, demand reacts on economic development. As people get more money to spend, they fly more often. In periods of economic downturn, they fly less. And when something bad happens, like a terrorist attack, a nuclear accident or an epidemic outbreak of a lethal disease, they don't fly at all. Aviation needs to plan capacity in these turbulent markets in such a way that aircraft always fly full, with airports that can handle the departing and arriving passengers efficiently. These market fluctuations are the basis for *tactical capacity planning*, day-to-day deployment of available capacity.

Aircraft need to fly all day to be profitable; airports are often closed overnight. Schedules are tight, but due to the technical complexity of aviation, the system is vulnerable to disruptions. A failing component can prevent an aircraft from flying, an incident on a runway can stop the operation of an airport and bad weather conditions can stop both. Due to the tight schedules necessary for profit, a small incident can disrupt these schedules causing delays. Recovering from these delays, bringing the system back to the planned schedules, is the basis for *operational capacity management.*

Capacity management in aviation is very much *data driven*. Data, such as on market demand, on economic developments, historic data and predictive data, are the main input for planning processes at the strategic, tactical and operational levels. *Statistics* are essential for interpretation of all these data.

*Modelling* these data and *simulating* various scenarios are essential in capacity planning at all levels. Planning capacity is complex, requiring robust planning software. Aviation operations are therefore very much IT driven.

## 1.5 Working in aviation

This textbook is intended for students seeking to enter the world of aviation and become one of the about 60 million people earning their living in this field. So, what does it mean to work in aviation? What does it need to become an aviation professional? What do the millions of employees in aviation actually do every day?

Young people with a keen interest in aviation often want to become a pilot, a flight attendant or an air traffic controller as these are visible jobs, jobs one can imagine from the outside. But most of the millions employed in this industry do other things. Most of the jobs can only be understood by clarifying how aviation functions.

Many people are fascinated by aviation because of its dynamic and international nature, as well as its glamorous appearance, as airlines and airports deliberately try to preserve the prestigious reputation of aviation in today's marketing communication. A student who wants to become part of this sector – an aviation professional – needs to look behind the romantic curtain as aviation is not an easy place to work. Not everyone is fit to work in aviation.

In the "old days" aviation was considered a sector for males only. In today's society, it is well recognized that aviation, as any sector, is a working place for males and females alike.

### 1.5.1 International environment

Aviation is known both for its attractiveness and for its harsh working conditions. This starts with the *English language* being the Esperanto of the sector. Anyone working in aviation needs to be proficient in the English language. Where most people perform their jobs using their mother language, the aviation professional needs to communicate in English. Understanding, speaking, reading and writing English is a must for any aviation professional.

The international nature of aviation also implies that we are confronted with time differences around the globe. This means that the aviation sector works round the clock, 24 hours a day, seven days a week. Obviously, flight crews on long-haul flights are directly confronted with these time differences, which can cause jet leg. But employees on the ground also need to cope as the 24/7 character of aviation makes *working in shifts*, even night shifts, a standard at many job positions, even in the higher-ranked jobs. Aviation professionals work outside normal office hours and may even work on Christmas Eve or Labor Day.

As aviation organizations work closely with other organizations all over the world, the ability to work together with people from other cultures makes working in aviation both fascinating and difficult. *Communication skills* are part of the basic abilities that any aviation professional needs to master.

### 1.5.2 Competitive environment

Aviation is highly competitive and the margins are narrow. Aviation fluctuates with the economy. Companies come and go; the border between profit and loss, between success and failure, can be very narrow. Working for a company on the edge of bankruptcy is not unusual in aviation. Alternatively, your company may be taken over by another company, making your job obsolete. In most aviation companies, *no job is for a lifetime*. Making a living in aviation implies changing jobs or changing employers at the right time to avoid unemployment. A dedication to *life-long learning* and keeping oneself up to date on developments is a prerequisite if one wants to be attractive to a future employer.

### 1.5.3 Professional competencies

The first and foremost competency that every aviation professional needs to possess is *self-criticism*. Aviation is unforgiving for mistakes or errors. Every aviation professional needs to bring a first-time-right attitude, whereby checking and double checking one's work output is part of normal behaviour. Colleagues need to be able to rely on each other's work output.

This brings us to the second-most important competency: the ability to work in a team as a true *team player*. It takes many people, many jobs, to operate aircraft and airports, to build aircraft or to maintain them. It's all team work, with no room for individuals.

Because of the complexity of aviation and its work with numbers – in the form of complex calculations, complex algorithms or complex planning schedules – an aviation professional needs a developed *quantitative mindset*. One must like working with numbers and calculations, and master the fundamentals of mathematics.

### 1.5.4 Level of education

There are only a few jobs in aviation that do not require higher education.

One can say that in designing and planning the future, professionals at the master's level are required. Designing aircraft and engines, designing new airports requires profound knowledge of technology or data. Thus, shaping the future of aviation and working at a strategic level requires a mastery of skills and competencies.

Operating existing equipment, be it aircraft or airports, creates jobs at a *bachelor's* level. Understanding technologies and planning methods, and applying these in an aviation organization form the jobs for bachelors in the aviation sector. At the execution level, where the work is done, most jobs require a firm educational background. As there is a permanent tendency towards more complexity, it is fair to say that many middle-level jobs have arisen or will grow towards a bachelor's level.

When it comes to managing aviation organizations we see both levels of education. Managing complex processes requires, next to educational skills, leadership qualities as well as experience. Recruitment of the higher ranks is more a matter of individual qualities than education level and is, by definition, primarily based on excellent performance in previous positions.

### 1.5.5 Engineering versus operations

Aviation can be classified into engineering and operations. Many aviation organizations, like suppliers, airlines or airports, have typical engineering departments apart from typical operational departments. *Engineering* deals with design, production, monitoring and maintaining technology, whereas *operations* deals with deploying that technology. Engineers need a thorough understanding of technologies applied in aviation, while operations need a broad understanding on how these technologies are effectively exploited. All need a profound knowledge of rules and regulations, and of safety and security.

In most job positions, the border between engineering and operations is vague, as both modes depend on each other. Engineering and operations are two sides of the same coin: while an engineer needs to understand the operational challenges, an operator needs to understand the technology that is applied. Both engineering and operational professionals are jointly responsible for operational readiness of aviation.

Most important of all, both need to understand their business environment. Everyone working in aviation needs to know and understand where the money comes from and how profit is made. Since money comes from customers, a high level of customer orientation is perhaps the most important quality required for a successful career in the service industry that aviation has become.

## Suggestions for further reading

Adam Pilarsky: *Why can't we make money in aviation*. Routledge, 2007.

Alan Dobson: *A history of international civil aviation*. Routledge, 2019.

Jose Sanchez-Alarcos: *Aviation and human factors*. Routledge, 2018.

Soren Eriksson, Harm-Jan Steenhuis: *The global aviation industry*. Routledge, 2009.

# CHAPTER 2

# Authorities in aviation

In the previous chapter, we saw that governmental authorities play many vital roles in aviation. Professionals need to understand these roles. Governments make *rules and regulations* that aviation players need to comply with; we will cover these in this book. Governments are also *enablers* of aviation, providing vital operational services, which we will discuss in this chapter

But let us first look at something very fundamental that nation states provide to their subjects: states define their *nationality*. Be it an individual person, an airline, an aircraft or its manufacturer, all entities have a nationality. In this book, when we discuss airports, aircraft and airlines, the nationality of these entities is the basis for all legal ruling.

Governments involved in aviation, despite their exclusive national jurisdiction, operate in *close collaboration* with each other. Rules and regulations, as well as the services provided by governmental bodies, are highly standardized on an international level, which makes aviation a truly internationally organized sector.

We will explore the notion of the nation state; then the concept of international lawmaking; and, lastly, the close cooperation between states in aviation.

## 2.1 Nation states

The world consists of 195 independent nations, *each of which is sovereign*, with full and autonomous control over its territory, territorial waters and airspace. We all live in this world and comply with its fundamental division into states. When visiting another country, we obey laws that might be different from the country we live in while visitors to our country need to comply with "our" laws. The exclusive *jurisdiction of the nation state*, the right to make and enforce laws within the own territory, is the basic modus of political power. This applies to almost everything in life, and certainly to aviation. *Every single nation state is sovereign in defining and enforcing rules and regulations in aviation.*

Now, as aviation is very international in its nature, this sector cannot operate under a patchwork of different rules. Therefore, *the need for cooperation* between independent nations was felt early in aviation.

Nations were used to cooperation in the *maritime* world. Still many basics in aviation regulation find their origin in maritime agreements.

Already in the first decade of aviation the first *multilateral aviation treaties* were made on fundamental elements like nationality and jurisdiction (the Paris Convention of 1919) or on liability (the Warsaw convention of 1924).

A multilateral treaty or convention is a formal document in which independent nation states transfer some of their independent powers to a jointly formed entity, a supranational body, whereby, thru the act of ratification, the underlying treaty becomes *part of the national law system*. The sovereignty of the nation state is guaranteed by every nation having a *veto right* on decisions to be made within the supranational body. Therefore, supranational bodies can only act by the *consent of all member states*. Even in the internationally oriented world of aviation, the sovereignty of the nation state is still the basis for power of making rules and regulations.

Enforcing such international rule-making is then executed by the nation state or, to be more precise, the state's *agencies*, like the police, customs, its aviation authority and its air traffic control organization.

With the exclusive jurisdiction of the nation state as the basis for political power comes the fundamental importance of *nationality*. Every entity – whether an individual person, an aircraft or an airline – needs to have an exclusive nationality. That nationality defines which laws that entity needs to comply with. An aircraft needs a *national registration* to become airworthy; we individuals all need a passport to travel abroad. These attributes define our nationality. It is very important to realize and remember forever that the international sector of aviation is built up of entities and parties with a nationality.

With the above in mind, we can understand the complex functioning of the supranational rule-making of ICAO and how this results in uniformed rule-making at national levels.

## 2.2 International cooperation: ICAO

The international dimension of states acting in aviation is formalized with their membership in ICAO. The International Civil Aviation Organization was formed in 1947 on the basis of the *Chicago Convention of 1944*, which is the *constitution of the commercial aviation sector*. ICAO is a specialized agency of the United Nations, like the WHO (World Health Organization) or the IMF (International Monetary Fund). One hundred and ninety-one member states of the UN are members of ICAO. All states are represented at ICAO's yearly *Assembly*. Every three years the Assembly elects 36 countries to form ICAO's *Council* located at ICAO's head quarter in Montreal, Canada. The Council prepares changes in rule-making that need to be enforced at the yearly assembly by *all states agreeing* as all states have a right to veto proposed changes.

The Chicago Convention stipulates *how nation states should treat each other* and sets standards on nationality, jurisdiction and commercial agreements between states, known as Air Transport Agreements. These ATAs define which airlines are allowed to operate to which countries and airports. We will discuss this in detail in Section 2.4.

The Chicago Convention sets the basis for *operational rule-making*, defined in detail by the so-called *Annexes* to the Convention of Chicago. There are 19 of these annexes, all dealing with specific elements of aviation. The operation of aircraft is dealt with in Annex 6, while airports around the world need to comply with Annex 14.

ICAO's annexes *are the basis for national legislation*. ICAO member states need to include the content of the annexes in their national aviation legislation. By doing so, all states have identical or at least comparable national aviation legislation.

Annexes by ICAO can contain both standards as well as recommendations. Standards, as defined in the annexes, are very strict; the member state needs to include these in their national legislation. Standards in the annexes use the words like "shall" or "must".

ICAO annexes may also contain *recommendations*, whereby the states need to follow these recommendations as a minimum standard in defining national aviation legislation. Recommendations use words like "should" or "may". These recommends are normally followed, but sometimes states divert from them for political reasons. When looking at how rule making by ICAO develops, it is fair to say that recommendations often become a standard after a certain amount of time.

The standards and recommendations following from the annexes of ICAO are referred to as SARPs, standards and recommended practices.

Hence, ICAO annexes form the basis for *national rule-making*. Nations impose national legislation on their airports, airlines and within their airspace, and this *national legislation is based on the 19 annexes of ICAO*. Every airport in the world complies with national legislation based on Annex 14, and every aircraft flies under a national regime based on the aircraft's registration, based on Annex 6.

ICAO does not possess powers to enforce rules; neither can ICAO sanction countries that violate ICAO ruling. *A state can only be sanctioned by other states*. If a country violates ICAO minimum standards, other countries may decide to suspend landing rights of aircraft

| | |
|---|---|
| Annex 1 | Personnel Licencing |
| Annex 2 | Rules of the Air |
| Annex 3 | Meteorological Service for International Air Navigation |
| Annex 4 | Aeronautical Charts |
| Annex 5 | Units of Measurement to be Used in Air and Ground Operations |
| Annex 6 | Operation of Aircraft |
| Annex 7 | Aircraft Nationality and Registration Marks |
| Annex 8 | Airworthiness of Aircraft |
| Annex 9 | Facilitation |
| Annex 10 | Aeronautical Communication |
| Annex 11 | Air Traffic Services |
| Annex 12 | Search and Rescue |
| Annex 13 | Aircraft Accident and Incident Investigation |
| Annex 14 | Aerodromes |
| Annex 15 | Aeronautical information Services |
| Annex 16 | Environmental Protection |
| Annex 17 | Security: Safeguarding International Civil Aviation Against Acts of Unlawful Interference |
| Annex 18 | The Safe Transport of Dangerous Goods by Air |
| Annex 19 | Safety Management |

**Figure 2.1** ICAO Annexes

registered in that country, or suspend operations to that country. As no country would want to be in such position, and as ICAO's minimum rules are meant to guarantee safe and efficient operations, it is in a country's *self-interest* to comply with ICAO standards. The only 'tool' ICAO has to influence countries to comply with ICAO Standards is the USOAP (Universal Safety Oversight Audit Program). The results of these audits are visible for all member States. By doing this ICAO uses 'naming and shaming' as a pressure tool to conform with the Standards.

In Annex 8, ICAO describes certification standards. These are not rules that stand by itself, but are the fundamentals that *National Aviation Authorities* need to follow. These national authorities are the entities that *issue certifications of airworthiness* and have the *power to enforce* the rules and regulations. ICAO set the standards, and national authorities make and enforce the rules.

## 2.3 National aviation authorities

Everything that deals with flying aircraft needs to be certified. This includes aircraft, their individual parts, the airports where they land or depart, the airlines that operate aircraft, maintenance of aircraft, flight crews and engineers. All need a *certification that declares them fit for flying* and compliant to all rules and regulations. *National Aviation Authorities* (NAAs), as part of the nation states, are responsible for issuing these certifications. They have the powers to enforce them, by *withdrawing* a certification in case of a non-compliance with the rules, by imposing legal *fines* to non-compliant parties or even *prosecution* of individuals violating the rules.

Formally all national aviation authorities are equal as the nation states that created them are equal in the society of nations. In real life, there is a clear *hierarchy* between authorities, as their position in the aviation world is formed by the scope of aviation activities within their respective nations.

As a certifying authority, the FAA, the US Federal Aviation Administrationand EASA, the European Aviation Safety Agency, are different from most other NAAs. This is because major hardware manufacturers in aviation possess US or EU *nationality*. On the basis of this fact, the FAA or EASA is the *designated agency* that certifies their equipment. The importance of being a designated agency causes the hierarchy between the NAA's; the nationalities of the leading suppliers in aviation. Therefore countries with important suppliers, big airlines and huge airports have aviation agencies that are more important to the process of rule-making than the agencies of countries that do not possess these big players within their nation. Certainly, every NAA is also a certifying authority, but for practical reasons, they primarily follow the agencies that originally certified that aircraft and that procedure. So, we see that it is the governmental agencies that make the rules and follow them.

The prime certifying agencies, therefore, are FAA and EASA (European Aviation Safety Agency). These two bodies certify most of the equipment as well as the methods of operating, repairing and maintaining that equipment. They also certify all the operators and their procedures as well as the level of education and training required. Since FAA and EASA certify most of aviation operation, they set the parameters for aviation design, production and operation.

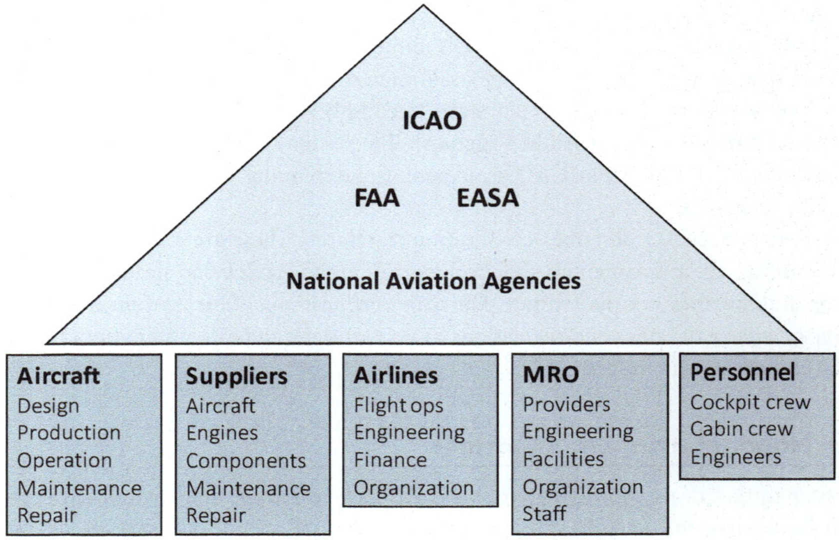


**Figure 2.2** The hierarchy of international rules and regulations

The aviation agencies of other countries where aviation suppliers reside, like Canada, Brazil, Japan, China and Russia, follow the FAA and EASA in their certification processes. They do so for a strong reason: they want to sell their equipment worldwide and need access to the US and the EU market. Therefore, their certification needs to comply with, and often even requires the involvement of, either the FAA or EASA.

### 2.3.1 FAA versus EASA

FAA stands for Federal Aviation Administration, the Federal Aviation agency of the *US Government*. EASA stands for European Aviation Safety Agency, the aviation rule-making agency of the *European Commission*, appointed by the 29 member states of the European Union, unifying the regulatory authority of the member states of the EU.

The FAA has been in place for a long time as it was felt in the US already in the early days of aviation that aviation is a *federal* affair. The states within the US have no independent jurisdiction on aviation. FAA not only creates the rules and regulations, *they also enforce them*. The FAA is a national aviation agency of a single nation.

EASA in its present form was created in 2003, after a period of unifying national legislation between the EU member states under the regime of JAR (Joint Aviation Requirements). EASA creates the rules and regulations, but *enforcement is the responsibility of the national aviation agencies* of each of the member states. In the UK, EASA regulation is enforced by the CAA, in France by the DGCA, in Germany by the LBA. All EU member states have their aviation agency, enforcing EASA rule-making.

The FAA and EASA cooperate to harmonize their legislation as much as possible. The outcome of this harmonization is one of the inputs for updating the ICAO Annexes. FAA certifies the Boeing 787, and is followed by EASA. EASA certifies the Airbus A350, and is followed by the FAA. This is possible as both certification processes are based on standards as defined by ICAO. FAA and EASA have a formal agreement in place for following each other's certifications.

## 2.4 Air transport agreements

As mentioned in Section 2.1, the Chicago Convention stipulates *how nation states should treat each other* and sets standards on nationality, jurisdiction and commercial agreements between states, called *Air Transport Agreements*. These ATAs define which airlines are allowed to operate at which airports. Landing rights for foreign airlines are granted by the state, and are formalized in *treaties* among states. Aircraft can only fly from one state to another state by consent of the state of departure, the state of arrival and the state of nationality of the aircraft. States need to ratify treaties with each other in order to enable international air services performed by airlines residing in the states. All states have ATAs in place between each other. These ATAs are the legal basis for international air transport. Most ATAs are bilateral: an agreement between two states. In Section 2.5 we will see that more and more these ATAs are becoming multilateral: agreements between more than two states.

The Convention of Chicago has defined principles of how an ATA should be structured, what it needs to contain and what freedoms states can grant to each other.

As such, an ATA can govern how many airlines of each state are granted to operate between the two states, how many destinations are allowed, what flight frequencies are allowed and which aircraft capacity may be deployed between the two states. If all of the above is strictly defined, we call it a *restrictive* agreement. If parts or all of the above elements are declared free, we call it a *liberal* agreement.

States can own their national carrier and can decide to *protect* the own flag carrier against competition from airlines of other countries. Such states agree on restrictive agreements. Restrictive ATAs are used by states to regulate air transport to and from their home airports in a way that protects the home carrier against others. Liberal states on the other hand let their airlines compete against other airlines on the basis of market forces. By creating liberal ATAs, states let their airlines perform commercially within a free market.

Most Air Transport Agreements are made upon the request of one or more national airlines that wish to operate to another country. However, as an ATA does not normally expose states to politically sensitive risks, the ATA is in the world of diplomatic relations between states sometimes used as a *diplomatic tool*. States can agree upon an ATA as a proof of recognition of the other state, or to express appreciation of actions taken by the other state. This means that there are "sleeping" ATAs, agreements that are in place but not used by any of the flag carriers between these states. Newly formed airlines, other than the flag carrier, can sometimes make use of such sleeping ATAs.

The freedoms that states grant each other are all defined and even numbered by the Convention of Chicago. They are known by the term The Freedoms of the Air.

### 2.4.1 Freedoms of the air

*The First Freedom* that states can grant each other is the right for an aircraft to fly in the airspace of the other nation, thereby overflying its territory, without the intention to land on an airport within that territory. In principle, most states have granted each other this right.

The first freedom imposes obligations to both states as well. The state where the aircraft is registered has the obligation to ensure that the overflying aircraft and its aircrew comply with all airworthiness rules and regulations. Furthermore, the overflying aircraft needs to obey all directions given by the ATC of the state that it overflies. That state in return has the obligation to provide the ATC guidance, in compliance with all rules and regulations governing ATC operations. Above all, the overflying state has to *ensure that its airspace is safe* for the overflying aircraft.

The First Freedom of the Air seems rather obvious and simple at first glance, but if we look at the mutual obligations it implies, it becomes clear that the First Freedom is actually crucial in the system of mutual trust between states; in the standardization of procedures; and, above all, for ensuring the safety of air transport.

*The Second Freedom* that states can grant each other is the right for an aircraft to fly in the airspace of the other state and to land on an airport within the other state for *flight technical reasons only*. The most common technical reason for such landing is to refuel the aircraft.

This second freedom has lost its importance over time as aircraft have considerably improved their range capabilities. Today's jet transports are able to fly non-stop to almost any destination, whereby the Second Freedom is no longer required. Cargo transports however still use it.

By granting each other the Second Freedom, states have the responsibility of ensuring the availability of an airport to land and that this airport meets all requirements defined by ICAO, as discussed in Chapter 3. Most important, granting the second freedom only makes sense if fuel is available at such airport.

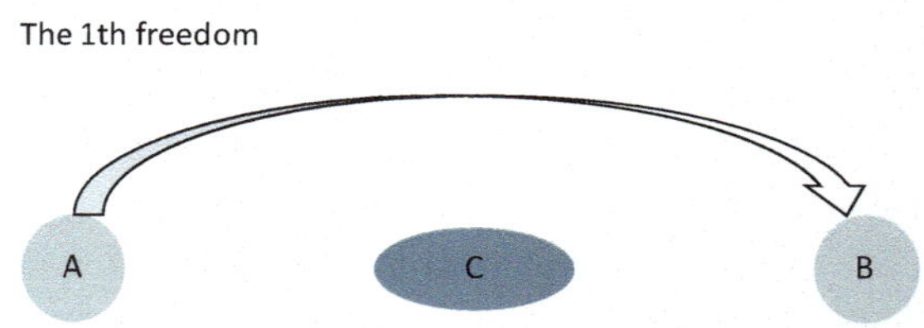


**Figure 2.3** The 1st freedom of the air

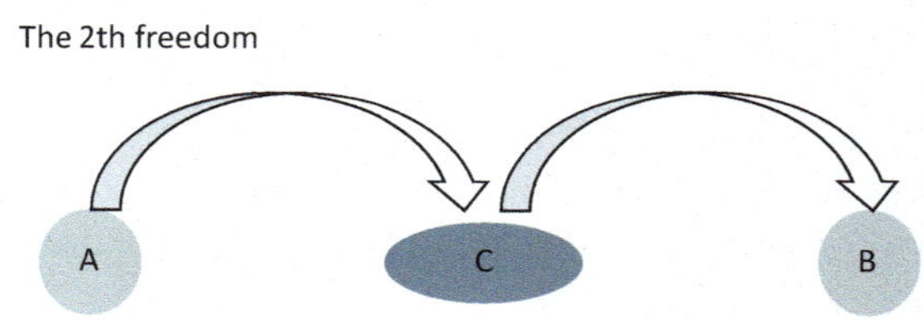


**Figure 2.4** The 2nd freedom of the air

The first two freedoms are also often referred to as *Air Transit Rights*. We noticed that most nations grant each other these rights, and this is formalized in the so-called *International Air Services Transit Agreement* – a multilateral treaty under the Chicago Convention. Some 129 states have ratified this treaty. But indeed, not all states have. States, like Canada or Russia, wish to remain free in deciding which aircraft they accept in their airspace. And this is not a coincidence; both countries possess massive landmasses in the far Northern hemisphere, airspaces that are needed by the airlines to operate long-haul routes.

*The Third Freedom* that states grant each other is the right for an aircraft to fly in the airspace of the other state, and to land on an airport within that state for *commercial* purposes. After landing, passengers and cargo disembark. The passage is paid for in the country of origin.

*The Fourth Freedom* is almost identical to the third, but now the passage can also be paid for in the country of destination. The third and the fourth freedoms form the basis for commercial air transport operation. Airlines fly between the two states, selling tickets in both countries, and compete against each other.

*The Fifth Freedom* is the right for an aircraft to fly from the home country to another country, land there for commercial reasons and continue the flight to a third country, by which the airline is granted the right to sell tickets for the passage between the country where the stop is made and the third country.

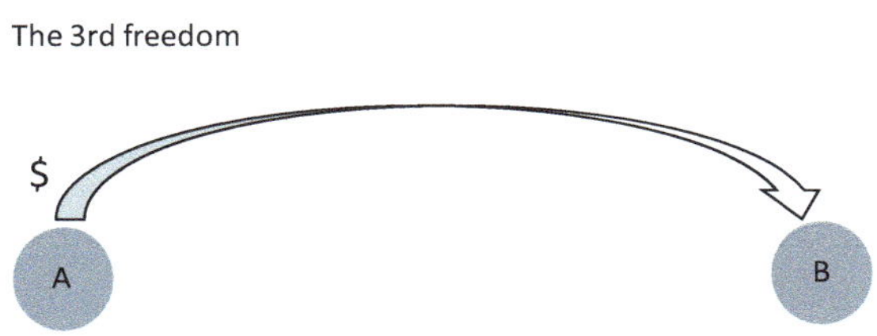


**Figure 2.5** The 3rd freedom of the air

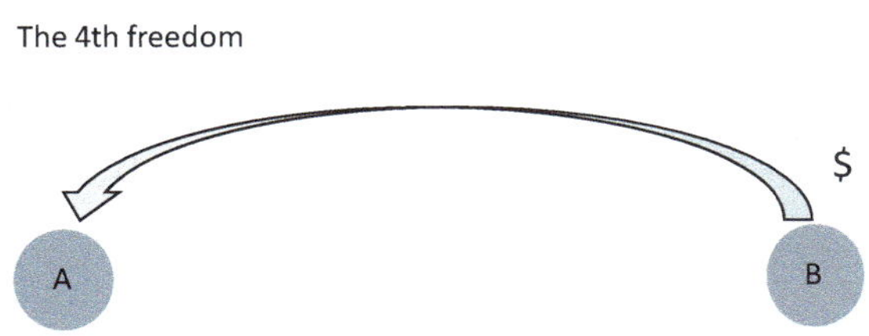


**Figure 2.6** The 4th freedom of the air

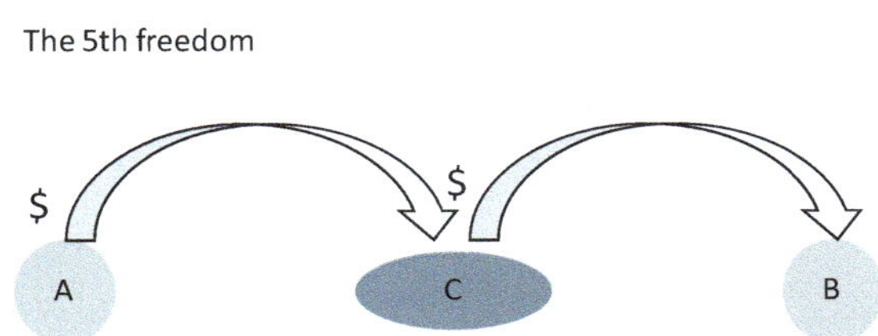


**Figure 2.7** The 5th freedom of the air

The fifth freedom is thus merely an extension of the second freedom, and the need for the fifth freedom indeed stems from the second freedom. In the age in which aircraft needed to land for refuelling, airlines wanted to exploit this technical stop by selling tickets to and from the intermediate stop, both with either the state of origin or the third state as the destination.

The fifth freedom is important for understanding the development of the airline business. Newly emerged airlines, formed in countries in the Far East and more recently in the Middle East, use the existence of ATAs containing the fifth freedom for their network expansion. These ATAs were agreed upon in the 1960s and 1970s, upon request by and benefiting airlines from Europe, the US and Japan. Airlines operating long-haul networks needed airports in the Far East and the Middle East for refuelling en route to a long-haul destination. As ATAs are by definition reciprocal and in principle never ending, newly emerged airlines now benefit equally from these ATAs.

*The Sixth Freedom* is the right for a national carrier to perform a flight from the home airport to another country and back, to perform a flight to a third country and back, and to sell tickets for passage between these two foreign countries with a transfer of flights at the home base.

This sixth freedom is the basis for today's hub airlines. Many airlines operate networks whereby all flights feed each other with passengers connecting at the airline's home base, the hub.

*The Seventh Freedom* is the right for a national carrier to perform commercial flights between two states other than the state of origin of that carrier. This right is seldom used.

*The Eighth Freedom* is the right for a national carrier to commercially operate form the home country to a destination in another country and then continue the operation to another destination within that country. The second leg of the operation is merely a domestic flight within the other state.

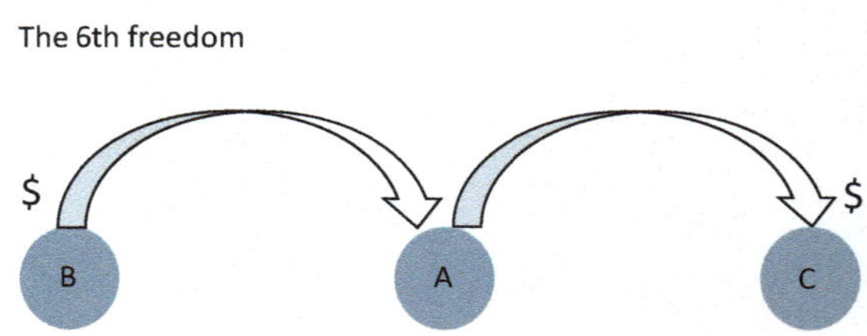


**Figure 2.8** The 6th freedom of the air

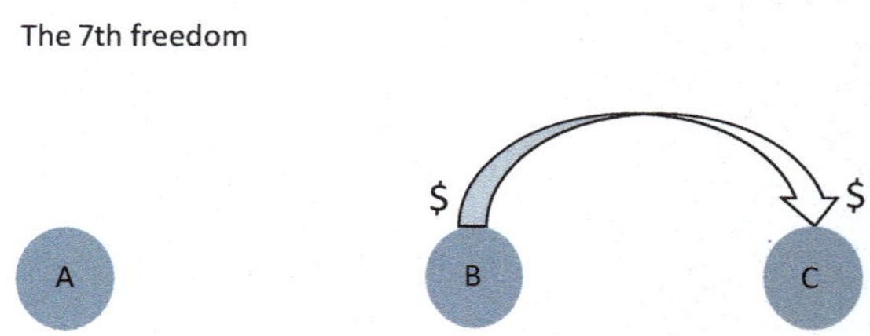


**Figure 2.9** The 7th freedom of the air

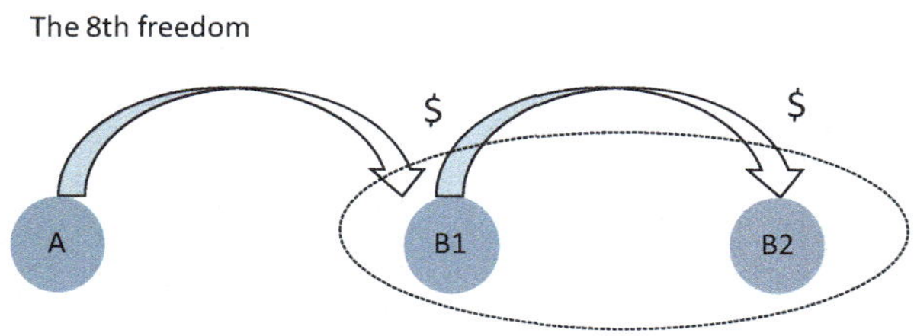


**Figure 2.10** The 8th freedom of the air

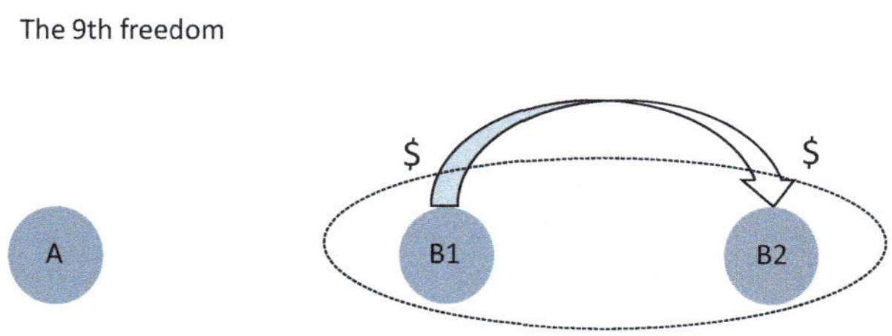


**Figure 2.11** The 9th freedom of the air

*The Ninth Freedom* is the right for an airline to operate between or to airports within another state so as to perform a wholly domestic operation in another state. This Ninth Freedom is also known under the term Cabotage, stemming from Sea Law.

## 2.5 Liberalization

As from the 1970s, with a steep increase in aircraft efficiency, aviation has developed into an economic sector based on *market demand*. Governments understood that aviation could develop better in a liberalized, market-oriented environment. The task of protecting national aviation vanished. Legacy flag carriers as well as new entrant airlines could become market leaders without protective support from their governments, but they could also go bankrupt without that protection. Aviation is now governed by *market forces*, and the process of gradual *withdrawal from commercial aviation by governments* is called the liberalization, or deregulation, of aviation.

This process is of *extreme importance* for understanding the growth and size of aviation, and what aviation is today. We tend to focus on the technological developments of cost-effective high capacity aircraft, but next to this technological development, the *political development of liberalization* is equally important for the massive growth of the aviation sector.

### 2.5.1 US

It started in the US with the *1978 Airline Deregulation Act*, stipulating that US airlines could fly from and to any US airport they wanted, against fares and service levels determined by the airlines themselves. This gave way to an explosive growth of the US domestic market.

Since 1992, the US has established so-called *Open Sky Agreements* with many countries around the world. Originally on a bilateral basis, the US also ratified multilateral agreements; an Open Sky Agreement is in place between the EU and the US. Any airline originated in the EU or in the US is allowed to operate from any airport to any destination in the opposite continent, with the only limitation that both airports are a declared international airport, with immigration and custom services available. EU airlines are *not* allowed to operate domestically within the US, *nor* do US airlines have the right to operate within the EU.

The US has bilateral open sky agreements with many countries around the world and is pursuing a multilateral agreement with the ASEAN countries.

### 2.5.2 EU

As from 1993, the European Union has declared that all airlines holding an AOC in any of the EU's member states have the right to perform flights between all airports within the EU. This means that EU member states grant each other's airlines all of the nine freedoms described above. There are no restrictions on either type of equipment or flight frequencies.

EU airlines are free to fly from any EU airport to any other EU airport, even if the two airports are within one state other than the state of origin of the airline. So, EU airlines can perform domestic flights in other EU countries, and indeed, some do.

As EU airlines have the right to fly to any airport they choose, neither the state nor the airport authorities can deny an EU airline from operating in that airport. The only limitation that the EU recognizes is the availability of a landing slot at an airport. If an airport is declared as being at the top of its capacity at a certain moment of the day, the slot coordinator of that airport can deny an EU carrier the ability to arrive at the airport at that congested moment. But here also, the EU seeks to find non-discriminative practices by declaring that in case of airport congestion, the available airport capacity should be fair and equally divided between the home carriers of that airport and the visiting carriers.

This entirely liberal free market for aviation within the EU has resulted in the development of airlines that operate within the EU from many home bases in many countries within the EU, and can do so on the basis of an AOC from any of the EU member states.

Hence, commercial aviation within the EU is governed by the liberal structure of internal EU ruling. For aviation agreements of the individual EU states with steps outside the EU, bilateral ATAs are still in place.

As many EU airlines felt the desire to expand the EU freedoms to countries outside, but in the vicinity of the EU, the EU has multilateral agreements in place between the EU member states and these third states. As such, multilateral agreements are in place with states that have *association agreements* with the EU. These countries are Switzerland and Norway, whereby EU airlines as well as Swiss and Norwegian airlines can operate as if their countries of origin were EU states. Swiss and Norwegian carriers enjoy the same freedoms as EU airlines within the EU, while EU airlines can operate to, from and within these countries as they like.

Comparable, but less liberal, are EU agreements with countries like Iceland, Morocco, Israel, Ukraine and Turkey. Here, airlines from these countries are allowed to fly from their home country to any destination airport within the EU, while EU airlines have the right to fly from the EU to any destination airport in these countries. However, airlines from these countries have no right to operate between airports within the EU, while EU airlines do not have the right to operate domestic operation within these countries.

### 2.5.3 Asia

In Asia, the ASEAN member countries have agreed upon a Single Aviation Market among the member states. This means that airlines in Asia can, with some exceptions, operate unrestricted to any international airport in the region.

China, not a member of ASEAN, bases its aviation policies on bilateral agreements, whereby the government is in control of the development of its Chinese carriers, and access to its aviation market by foreign carriers, and first and second freedoms, are based on bilateral agreements.

The same applies to Russia, where the government remains in control over its aviation sector. Due to the massive size and northern location of Russia, first and second freedoms are also based on bilateral agreements.

### 2.5.4 Middle East

At the moment of writing, the EU is engaged in formulating a multilateral agreement with the Middle East states, governing all airline operations between the EU and the Middle East, while liberal agreements with Asian states are on the agenda. The US has an open-sky agreement in place with states in the Middle East.

It is clear that with aviation becoming mature, with airlines that can successfully compete in a free market, the need for state protection is disappearing. The result is that restrictive bilateral agreements become less relevant, with the global aviation business becoming more and more governed by liberal multilateral agreements, allowing airlines and airports to act and develop in a free and liberal marketplace on a commercial basis.

## 2.6 Air navigational service providers

Governmental aviation organizations such as the NAAs not only set and enforce rules and regulations; they also provide services to aviation.

When looking at the services provided, we see that most of these activities take place at airports. We will therefore discuss them in Chapter 7.

A special service that the authorities need to provide is managing all air traffic in the airspace that belongs to the nation state. Here we see the state emerging as the Air Navigational Service Provider (ANSP). Every state is obliged to have an Air Transport Management organization that executes the national jurisdiction of the country in its airspace, either in a primary role or secondary to the military.

### 2.6.1 Air traffic management

Every nation needs to have an Air Traffic Management Organization. This ATM organization can be imbedded in the NAA, as is the case in the US, or can be a separate organization

Communication between ATC and the pilots of flying aircraft is done by *radio contact*. This communication therefore is *verbal*, and under ICAO ruling this verbal communication needs to be done in the *English language*. As radio contact can be disturbed, aviation has adopted a variety of communication rules to ensure that pilots and ATC controllers understand each other. To prevent ambiguity and misunderstanding ICAO has established a standardized phraseology in English that includes proper spelling (the ICAO alphabet Alpha, Bravo, Charlie), proper pronunciation of numbers and digits and the use of narrowly defined 'code words' like 'affirm' for 'yes', 'negative' for 'no', 'roger' for 'message received'. All aircraft and all ATC stations have a *call sign*, whereby the aircraft's call sign is composed of airline coding and its flight number. Large aircraft, generating wake turbulence, have the addition "Heavy" in their call-sign, for the A-380 it is even "Super". In order to ensure that the pilot and the ATC controller have understood each other, all information given by one is acknowledged by the other.

Aviation operations have become very digital; digital development also enables digital communication between the pilot and the controller. This development is called CPDLC, *Controller-Pilot Data Link Communication*, and is gradually implemented. It requires sophisticated digital equipment both on the ground as well as in the aircraft. Modern aircraft are equipped with such digital data communication systems allowing CPDLC.

Surveillance and monitoring of aircraft by ATC is primarily done by means of radar and, more importantly, *secondary surveillance radar*, SSR, where basic, or primary, radar is dependent on processing of the reflection (or 'echo') of radio signals from the aircraft. Secondary radar is dependent on the actively sent radio-reply from the aircraft on a request from the (radar) ground-station, and includes information of the aircraft, like call signs, altitude, speed and track.

Presently the aviation industry is in the process of introducing ADS-B, Automatic Dependent Surveillance-Broadcast, under EASA in 2017 and FAA in 2020. This device, installed in aircraft, determines the aircraft's position by means of GPS and broadcasts its position, together with information on the aircraft's speed, heading and altitude, to any other ADS-B user, be it other aircraft, website or ground station. This technology will in due course replace today's ATC.

ATC is a complex organization, involving the inputs of many highly educated employees, supported by sophisticated equipment. ATC normally consists of three interdependent departments: air traffic control on the ground as well as in the direct vicinity of an airport for taxiing, take-off and landing; air traffic control in the upper airspace; and air traffic control in the 'merge area' (or terminal manoeuvring area), where traffic is directed from the upper airspace to an aerodrome and vice versa.

As these two departments are directly related to the operations of the airport, we will deal with these in Chapter 9. The third department is Upper Airspace Control.

### 2.6.2 Upper airspace control

Upon leaving the airport's terminal area, the departing aircraft is declared en route, and is handed over to the ATC's UIR, *Upper Information Region*. This part of ATC governs all aircraft en route and hence also aircraft that pass a nation's airspace at cruising altitude without the intention to land.

to take action if such a situation – the seizure of an aircraft – has occurred, be it confirmed or suspected. If a suspected aircraft does not respond to the verbal radio contact initiated by UIC Control, this part of ATC has the authority to hand over this event to the military. The Air Force of the state will attempt to make physical contact with fighter planes. If an unidentified aircraft flies into a forbidden corridor, this event even may lead to shooting down the suspected aircraft. ICAO provides the rules of engagements here.

As mentioned before, airways within an FIR go up to approx. 47,000 ft. Aircraft capable of flying higher, like many of today's executive jets, often get the freedom to fly their preferred route. This is called *Free Flight.* At these higher airspaces traffic is thin, and aircraft today are all equipped with TCAS Traffic Collision Avoidance System. This is a computer connected to the transponder. It detects transponders of other aircraft in its vicinity, and even communicates with these in case of a potential collision course with that transponder.

Every country has one or more FIRs that define the controlled airspace above their territory. So, what about the open sea, where no country has exclusive jurisdiction? Also above the High Seas, FIRs are defined and controlled by the UIC Control of the coast state. Above the seas there is normally no radar coverage. Here UIC Control uses the so-called Procedural Control. In short, this means that an aircraft upon entering an airway over the sea obtains a fixed flight level and a 15 minutes' horizontal separation with other aircraft.

### 2.6.3 ATM cooperation

As said, almost every country has a separate FIR, and large countries have more than one. Therefore, an aircraft performing a flight needs to pass many FIRs, whereby all ATC organizations are involved in composing and executing a flight plan. This makes flying through the airspace complicated and expensive. Navigation charges form a sizable part of the expenses of airlines. The need to fly designated routes or airways in every single individual FIR causes problems. When the need to follow routes or airways and to pass designated checkpoints en route is removed, each aircraft may fly their individual preferred route and altitude. This is called Free Flight. This works of course best in vast blocks of airspace governed by a single ATM-provider. For this reason, the aviation sector asks the governments for rationalization, by reducing the number of FIRs. This has led to the forming of Eurocontrol, and the drive towards SES in Europe, as well as Next Gen in the US.

Founded in 1960, Eurocontrol is an international organization with 41 European member states. The general objectives of Eurocontrol are to standardize ATC procedures in the upper airspace over Europe, and to coordinate the operations between civil ATC and military ATC.

Most notable result of Eurocontrol is the formation of a single operational FIR in the upper airspace above FL245 of Belgium, Luxembourg, the Netherlands and northwest Germany. This joint airspace operation is referred to as MUAC, Maastricht Upper Area Control Centre. Eurocontrol is also the designated organization that prepares the European ATM community for SES.

SES stands for *Single European Sky.* The aim is to have a streamlined and efficient European ATC in the upper airspace in operation by 2020. The airspace over Europe today is divided in 61 FIRs. Under SES this will change to a division of that airspace in 9 FABs,

Annex 13 clearly defines how the states involved should inform each other, whereby all states involved are entitled to be informed and have the obligation to inform the others as well.

## Suggestions for further reading

Benjamin Scott, Andrea Trimarchi: *Fundamentals of international aviation law and policy.* Routledge, 2019.

Pablo Mendes de Leon: *Introduction to air law.* Kluwer Law International, 2017.

CHAPTER 3

# Structure of aviation supply

This chapter is about the suppliers in aviation: the *firms and companies* that deliver products and services in aviation with the purpose of *earning profits*. It also introduces the fundamentals of *business administration* in aviation. In many ways, aviation is a high-value global business like others, but in some key elements, it is a very special kind of business, which we will concentrate on in this chapter.

The aviation supply side is a difficult sector to manage and work in as managing and operating aviation companies is complex and unforgiving of mistakes, and involves job insecurities for the workers. Developing products and services require *huge investments*, which are to be earned back in the *long-term future*. This future is uncertain and changes constantly. As many companies provide comparable products and services, *competition* is fierce, both on price and on product quality.

In the constant rivalry amongst companies in aviation, the winner takes it all. There is some room for a second place, but hardly for a third.

## 3.1 Ownership of aviation suppliers

Ownership of aviation suppliers can differ throughout the world. In many countries aviation companies are privately owned, in other places aviation companies are owned by the state. But also at privately owned companies, state involvement is sometimes not far away. Privately owned companies can have a strong need for governmental involvement in order to limit financial risks on product developments that are regarded as strategic by governments. The strategic consideration for governments can be for innovative technology development, but can also be for industry politics, with an emphasis on labour employment. For large states the main strategic consideration is often defence as the aviation business provides air power, which is the defence system's *most important* weapon.

The relation between ownership and nationality is also a complex one in the aviation business. In Chapter 2, we saw that every aviation player needs to have a nationality. In the case of airlines, the majority of owners need to have the same nationality as the airline. In case of airports this is not necessary, and many airports are owned by investment entities like pension funds residing in other countries than the airport itself. Key consideration for the investors here is a stable long-term return on the investment. Airport investors own the real estate of the airport, and expect some dividends from the operational profits of the airport.

Most suppliers are privately owned. This can mean that their stocks are traded on the stock exchange, but can also mean that a company is partly or wholly owned by investment funds. As aviation companies often have huge turnovers, profitability in absolute numbers can also be substantial as margins vary widely within the aviation business. In today's environment, many aviation companies are owned by other, larger aviation companies. The key word here is *consolidation* and indeed we see the aviation business becoming the playing field of a limited number of conglomerates.

## 3.2 Margins and profitability in aviation

One of the fundamentals of the aviation business is the wide *variation* in profitability between the various sub-sectors of the aviation business. We see aircraft and engine producers struggling against low margins, while specific component suppliers enjoy high margins. We see airlines operating at low or at best moderate margins, and airports sometimes being very profitable. Aviation is a business of *inequalities*. Even more if we realize that the NAAs that we discussed in Chapter 2 are not concerned with profitability at all; they can charge the industry as they see fit within their monopolistic positions.

For the suppliers, it all boils down to the amount of investments required for developing new products. For airframes and engine manufacturers the required investments are enormous. The calculation is in terms of billions, not millions. These investments need to be returned over time by selling the products for good prices. But as various suppliers offer comparable products, these prices are constantly under pressure. As a result the market prices for aircraft and engines may cover the cost of production, but they hardly cover the cost of development.

Aviation suppliers often try to recover from low selling prices by charging high prices for parts and services once the equipment is in operation. But here they are confronted with a wide variety of independent service providers who are able to offer lower prices due to their lower cost base.

Suppliers of aircraft *components* can often do much better. Their equipment is often fitted in various aircraft types, and due to the technical complexity of their components they can often protect their aftermarket sales against independents. Besides that, if a component fails to the extent that the aircraft becomes AOG, the supplier of the component has a strong negotiation advantage in setting the price for a replacement part.

For airlines, the bulk of the cost is labour, fuel, owning the aircraft itself and the charges that service providers impose on the airline's operation. Here we see that airlines originating in countries with *low labour cost* have the advantage. As most other costs are either comparable or equal, the *regionally different cost for labour* is a prime discriminator for the cost differences between airlines.

Many aviation companies have *complex organizations*. Developing and manufacturing products require many people working in a huge organization, and supporting products

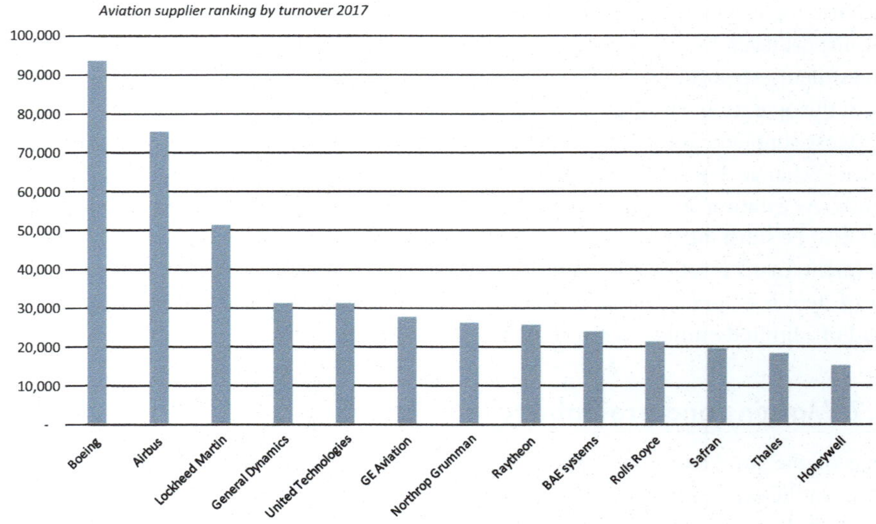


**Figure 3.1** Aviation supplier ranking by turnover

require a global organization for staying close to the end user of the products. Airlines also employ many people across many departments, resulting in complex organizations. Here we see that differences in managing these complexities lead to differences in *overhead costs*, resulting in huge differences in profitability between aviation companies.

On a more general level, we see in aviation a difference in *the need for profitability*. Privately owned enterprises need to be profitable for their owners, whereas for state-owned enterprises, this need for profitability is less urgent. It all boils down to the question: what do the owner(s) want? Even in a very market-oriented society like the US, many airports are owned by the municipality. The airport is regarded as vital infrastructure for the city or region and does not need to be profitable by itself. Airports that are privately owned are required to be profitable. A state may exploit an airline for many reasons, whereby profitability is not always the paramount driver. Thus, there is widespread need for profitability among aviation organizations.

## 3.3 The need for economies of scale

In order to ensure cost efficiency, many aviation companies seek economies of scale. The idea is that the bigger the company in terms of production output, the lower the total cost per product.

This need for economies of scale has really shaped the business landscape of aviation. Aviation suppliers have been in a constant process of consolidation. In all aviation producing countries, a multitude of companies have been merged or been taken over into fewer and fewer companies. In today's aviation business, only a handful of huge providers have remained.

The process of *consolidation* can come in various forms. First, and mostly applicable to US companies, are large companies buying smaller companies. That can be a smaller direct competitor (*horizontal concentration*) or a supplier of equipment that is related to the products of the buying company (*vertical concentration*). This large buying company can decide to let the smaller company continue its operation under its original name, as is done by UTC, or under the name of the buying company as is done by Honeywell. The buying company can also decide to terminate the company that they bought. This is what Boeing did after taking over McDonnell Douglas in 2000.

In Europe, the consolidation process has more been a process of mergers, whereby first on national scale suppliers merged into single national companies like British Aerospace and Aerospatiale. After the national consolidation came the supranational consolidation via EADS finally into Airbus.

In the airline business, the need for scale is also recognized, but the *need for a single nationality* has prevented airline consolidation to happen on a multi-national scale. Only within large aviation countries like the US, domestic consolidation has taken place. Also within the EU, where all airlines have the same EU-nationality, we see airline consolidation occurring, with British Airways buying the Spanish flag carrier Iberia and Air France taking over the Dutch flag carrier KLM.

The second reason why consolidation in the airline industry did not take place at the supplier side is the fact that it is relatively *easy to start a new airline*. And indeed, the airline business is shaped by newcomers, with important airlines that did not exist in the past. New entrants can bring with them new ways of aviation operations. In the airline business, we see new airlines lowering their cost by simplifying their operation, the so-called Low Cost Carriers. But if we focus on this category of airlines, we see that by accepting low margins, these airlines also seek for scale in order to ensure profitability. Therefore, in the airline business we still see a wide variety of airlines, but individual airlines seek to grow by expanding the network, by taking over smaller airlines or by forming alliances or joint ventures with other airlines.

## 3.4 An oligopolistic business

The need for economies of scale within the manufacturing part of aviation business has led to only a handful of providers remaining in the aviation business today. And this is truly remarkable: a global business transporting some four billion passengers per year relying on only a handful of aircraft manufacturers and engine suppliers. One could easily think that these suppliers can set their prices as often is the case in oligopolistic markets. But this is not the case in aviation, as the handful of providers mostly offer comparable and thus competing products, resulting in sometimes bloody price battles between them.

Although very limited in setting the prices, the same handful of aviation suppliers define the *structure of the aviation business* to a large extent. This is due to the simple reason that these suppliers develop the technologies that are applied in the business. By applying emerging technologies, individual suppliers can really change the business, and successful technologies are quickly copied by the other suppliers. As such it is fair to say that the combination of the technological complexity of aviation products on one hand, and the oligopolistic

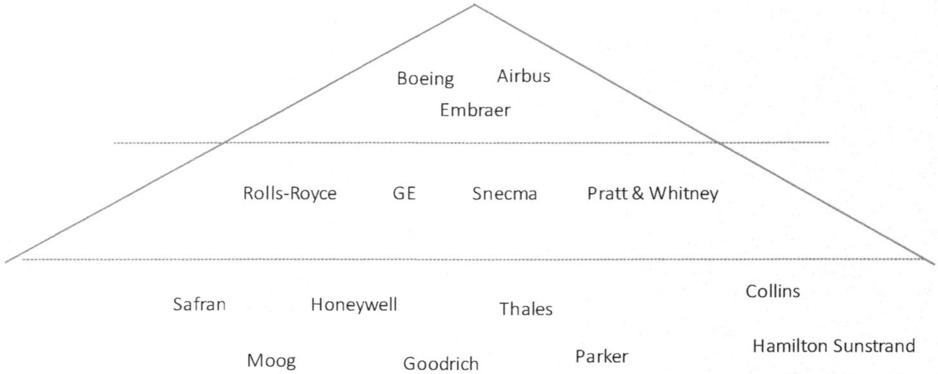


**Figure 3.2** Supplier hierarchy

structure of the industry on the other hand, place the suppliers of aircraft, engines and components in *a leading role* in the development of the aviation sector. All airlines operate the equipment that they develop, and all passengers fly in the aircraft that they produce.

## 3.5 Link to the military and to space

This textbook is about commercial aviation, with insights into the suppliers of commercial aviation. However, in order to understand these suppliers, it is worthwhile to realize that many of them, certainly the larger ones, are also engaged in military projects and space programmes.

Here the business structure is slightly different. Military projects often have less market pressures, and price setting is often based on cost plus a margin. Military projects therefore can be very profitable to suppliers. And indeed, for many large suppliers it can be said that the majority of *turnover* is in commercial aviation, but the largest part of the *profit* is in the military projects. Military projects can become even more lucrative when technologies developed for a military project funded by the government can be deployed in commercial aviation projects. This is even more the case in space projects. In space, mission capability and reliability are more important than cost. Space projects therefore can greatly contribute to a company's technical capabilities and financial health, two qualities necessary for developing commercial products.

## 3.6 Business-to-business markets

Perhaps the greatest challenge for suppliers in commercial aviation is their task of translating technological developments into value propositions to their customers. Suppliers are very technology driven, and are constantly developing new technologies. But while doing so, suppliers need to answer the following question: what will the new technology bring to the market, and how can customers buying that technology earn profits by deploying that new technology? This is far from simple, and mistakes can lead to catastrophic results. Perhaps the most striking mistake in the development of commercial aviation is the technology-driven quest for supersonic speed, resulting in the Concorde. Fast as it flew, no airline would ever be able to earn a profit with this technologically magnificent machine.

Successful application of new technologies concentrates on the equipment's fitness for use, operational reliability and reduction of manpower requirements. Most of all, technological developments are used for improving aircraft fuel efficiency. Lowering the fuel bill is not the primary key driver here. Lower fuel consumption means more payload and more range, with lower noise and pollution around airports. And more recently urgent, less fuel means less $CO_2$ emissions.

This all implies that the suppliers need to understand their customers, the airlines, the airports and the National Air Service Providers. With this notion, the vertical organization of the aviation business becomes clear: suppliers sell their products and services to airlines, and airlines sell their transport services to passengers and freight forwarders. In terms of money, the flow goes in the opposite direction: passengers buy tickets with airlines, and with that money the airlines pay the suppliers. Understanding this basic principle is a prerequisite for understanding aviation as a business.

## Suggestions for further reading

Daniel Todd, Jamie Simpson: *The world aircraft industry*. Routledge, 2013.

Herbert Baum, Stefan Auerbach: *Strategic management in the aviation industry*. Routledge, 2016.

## CHAPTER 4

# Structure of aviation demand

This chapter may well be regarded as the key chapter of this book. We could have started with it; actually, we did by observing that aviation has become a *demand-driven industry*. Demand forms the *income stream* of aviation. This multi-billion industry is foremost fed by income from passengers and freight forwarders paying for transportation. For workers in any of the aviation enablers or providers the same rule applies: *the customer pays your salary*, which is all the more reason to have a closer look at demand.

Demand for aviation comes in many forms, with different types of passengers and freight. All have one common denominator: all demand in aviation is *derived* demand.

This means that demand for aviation comes from demand for persons or goods to be transported to another place. Demand for aviation is not an autonomous demand as for food or a TV-set. The notion of aviation demand being a derived demand has some far-reaching complications for understanding demand for aviation.

First complication is in the possibility of *substitution*: the demand for transport can be fulfilled in various ways. For short distances, ground transport like the train or the car is a viable substitute; for long distances, electronic devices like the conference-video call can be an alternative to physical presence. Demand for cargo transportation can also be fulfilled by ship, lorry or train.

A second complication is that demand for transportation is *volatile* and reacts to changing conditions. Demand for transportation to a popular holiday destination can drop as a result of a natural disaster in that country, and demand for transportation of goods can change due to relocation of the assembly lines or the development of new markets at other places.

A third complication is the limited marketing techniques with which to persuade customers to buy. Where other industries, like food or textiles, can create an autonomous demand for their products, aviation needs *to follow the demand for transportation.*

Most of all, the derived demand for aviation differs per person or good, depending on the reason for the demand for transportation. Some passengers *need* to fly, many *want* to fly and most passengers *like* to fly. This need, want or like defines the *price elasticity of demand* for air travel – the rate at which demand reacts on price. A low elasticity of demand, as for food or water, means that consumption will not react fast to price changes. On the other hand, when this elasticity is high, as it is on luxury products, demand will react quickly to price changes.

## 4.1 Passengers

Regarding the demand for air transportation for passengers, we need to look at the *propensity to fly* by that passenger. This contains two elements: a *demand* for transport and the *financial capability* to pay for air transport.

Demand for transport comes in many forms. Originally it came from international business and from tourism. With economic development, this demand grows. But economic development is very unequally spread around the globe, resulting in mass migration not just on a continental scale but on a global scale also. This migration has created massive markets for aviation, both continental and intercontinental.

With economic growth comes a new middle class wanting to travel and an increased ability to pay for air transport. On the supply side, this transportation lowered in price. This combination is boosting the propensity to fly. Here we see the ingredients for the impressive growth in aviation over the last decades.

Demand for air transport is massive today, and the actual demand for an individual passenger can be based on a variety of reasons, making demand segmentation more difficult. The more a society integrates and develops economically, the more reasons there will be for travel up to a point where air transport is a normal and indispensable travel mode, as in the US domestic market. In such markets passengers fly as they want, and this is a basis for market *saturation*. With the economy going up, such markets do not grow to the same extent, and when the economy goes down, it does not shrink accordingly either.

Nevertheless, it is possible and necessary to segment passengers into three main categories: the business traveller, the VFR traveller and the leisure traveller. This is helpful for planning purposes as most passengers behave according to one of these segments. As said, demand for air travel is a derived demand for transport, and as such we can distinguish between passengers who need to fly, passengers who want to fly and passengers who like to fly.

### 4.1.1 The business passenger

This is the passenger who needs to fly for business or for work. This passenger travels very often and to business destinations. As this passenger needs to travel, the *price elasticity* for this passenger is low; he, or she, will travel, irrespective of price. Flights are selected on the basis of schedule *adaption to the agenda* of the traveller, and on total *travel time*. The business traveller prefers flying *non-stop*. As a *regular* passenger, he or she has experience and has preferences regarding airlines or transfer airports. This passenger often flies in the *higher booking classes*; airlines therefore call that ‘business class’. Business travellers flying often in the highest class generate sizable income streams for aviation.

Business travellers prefer air transport over ground transport already at fairly short distances; speed is of the essence, and for shorter destinations the business traveller wants a *same-day return*, avoiding unnecessary stays in hotels.

### 4.1.2 The EVFR passenger

This is the passenger who wants to fly. VFR stands for *visiting friends or relatives*; indeed here we have a passenger who is part of a migrant family. Families reunite on holidays, like Christmas, Chinese New Year or Diwali, or for marriages and funerals, resulting in massive markets for air transport. With unequal economic developments on a continental scale, like China or the EU, comes the migrant worker. Born, raised and educated in a poor region or country, people move to the richer regions or countries for work, resulting in vast "*employed-abroad*" markets. This passenger behaves like a VFR traveller, so combined they form the EVFR segment.

Behaviours of this passenger are formed by the notion that he or she always travels to the *same destination* and on a regular basis. He or she knows what is available in the market, both in airline quality and in airport quality. This passenger is price conscious, but seeks a price/quality balance and is prepared to pay for what he likes most: non-stop travelling or a transfer at the favourite airport. The *price elasticity in this market is medium*, and in case of emergencies in the family, demand becomes inelastic. The EVFR markets are seasonal, as they come with special days like Christmas or Diwali, but outside these peaks demand is steady available as well.

### 4.1.3 The leisure passenger

This is the passenger who likes to fly, visiting faraway countries and cities for holiday. Travelling is a luxury product and the *price elasticity is high*. When it is considered as too expensive this traveller will simply choose another destination, another transport mode or will not travel at all. Luxury products are the first to economize on when necessary. This makes the leisure market a very big but *volatile* market. Above all, this market is very *seasonal*. Contrary to the first two sorts of passengers, this passenger also need transport and accommodation at his destination, many leisure passengers buy an all-inclusive holiday package, of which air transport is just a part. This passenger travels incidental and mostly to different places. As most holiday travellers seek the sun and the beach, destinations

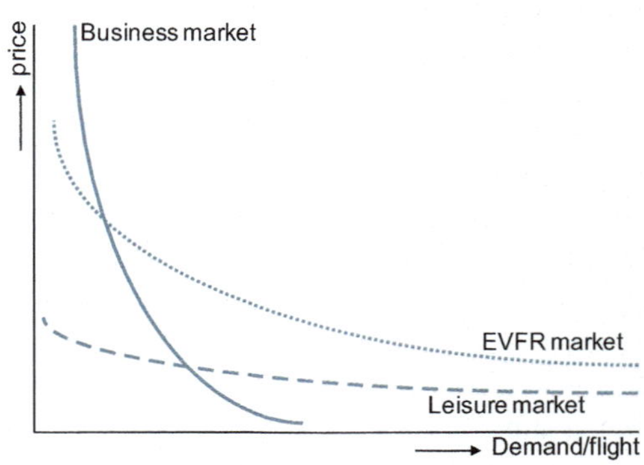


**Figure 4.1** Price elasticity of demand

are interchangeable. The weekend package to a popular city has become a commodity in which marketing techniques, especially from the food industry, such as day discounts or impulse purchases, have become effective. In all cases, the price is key, and most go for the lowest price. Margins in leisure business are marginal and high production volumes are necessary for low production cost.

### 4.1.4 Social demand

Apart from the three sorts of passengers described above, we also see social demand for air travel. This is in low populated vast areas often with harsh climate conditions and low ground transport availability, or islands belonging to the mainland. Here aviation may well be the only way of public transport. This air transport is often subsidized by the national or local government in order to keep fares affordable at each level.

### 4.1.5 Military demand

Large countries, or countries with military involvement in other parts of the world, can generate demand for *troop transport*. Troops can be transported in passenger aircraft, and the military branches of such countries may well sort-out this transport to commercial airlines. This military demand can be temporary, in the case of a military crisis, or structural, when the military presence in another part of the world is permanent.

## 4.2 Freight

Aviation transports not only passengers, but also freight.

Demand for transportation of freight is directly linked to economic patterns for trade and production of goods. Products are produced in other places than where these goods are sold, food may be raised or caught at other places than where it is consumed. In terms of weight, only about 1% of world trade is transported by air, but in terms of money values this is about 35%. Aviation is indeed an indispensable element of global logistics.

### 4.2.1 Cargo

Most of the transport of freight is cargo; goods in the form of final products transported to the markets or semi-finished products transported to assembling facilities. The main reason for choosing air transport is the *short travel time*. This is relevant for *perishables* like fish and flowers that lose their market value quickly. It is also important for valuable goods, like consumer electronics. Here the speed of transportation results in managing the current assets on the balance sheet of the owner of the goods, so the quest for speed is *financial*. For the textile business speed can be essential for *seasonal* clothing to be brought to the markets at the right moment. *Life stock* also is prone to long transportation times as it needs to be kept under special conditions, and for long range distances a ship is not a viable alternative. This also applies to pharmaceuticals as air transport offers the best guarantee of meeting climate conditions during the transportation.

A special advantage of air transport is in security of such transport. Valuable paintings or nuclear material is flown from and to the nearest airport, optimizing the security of such vulnerable and valuable transport.

### 4.2.2 Mail

The oldest demand for aviation has been mail, due to its speed. In quantity mail is just a fraction of the total demand for freight, but against high money values. Mail can have the form of documents like letters, or packages. Mail transport on longer distances faces increasing competition form parcel operators.

### 4.2.3 Parcels

A special form of freight demand is in demand for transport of parcels. A parcel can be an individual product sold at another place than where the buyer resides. Often, a parcel is a document that may need to be signed in hard copy somewhere else than where the document was produced. The fast proliferation of internet sales on a worldwide scale generates a vast demand for parcel transportation. Here also, speed is of the essence.

## 4.3 Demand variation

Demand for aviation comes in many forms and at increasing quantities. An overwhelming challenge in fulfilling this demand is that the actual demand varies constantly on a day-to-day basis. Production capacity is much more fixed on a time scale; thus we see instances of peaks where demand outnumbers the available capacity, and moments where capacity outnumbers actual demand. It is the constant variation in demand that makes capacity management in aviation challenging.

### 4.3.1 Fluctuation of demand

The first form of variation is in the fluctuation of demand. Actual demand fluctuates per season, per week and even on a per day basis, all due to the derived need for air transport.

Seasonal fluctuation is mainly generated by the leisure travellers who want to fly during the holiday season. This may be during summers, but it can also be during winters for Northern countries. Outside the typical holiday periods, leisure demand is weak.

In terms of long haul, we see high demand early in the week and at the end of the weak as many long-haul business travellers typically spend a work week at the business destination. All want to be home for the weekend.

For the day-return, we see in continental markets a peak in the early morning and early evening, with low demand in between.

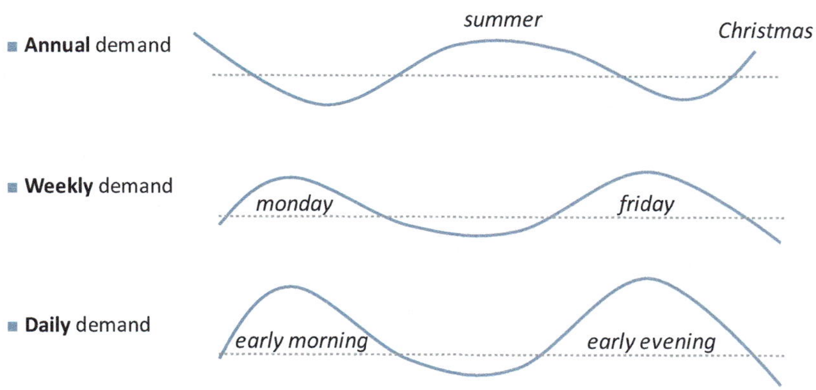


**Figure 4.2** Fluctuations in demand

## 4.3.2 Imbalance of demand

When demand for an outbound flight differs from the demand for the return flight to the home base, we have imbalance of demand. This imbalance is partly the result of fluctuation. The early morning flight to the business destination is at high demand, whereas the return flight around noon faces low demand. The fully packed long-haul flight departing on Sunday afternoon may fly back half empty on Monday. In Part IV we will discuss airline hub systems, and here we will see that feeder flights – flights that feed the hub – often have an imbalanced demand pattern.

Freight demand is imbalanced by definition. Here we can assume that most passengers buy return tickets, whereas freight goes one way by definition. This implies that demand for a certain route can only be fulfilled optimally if a return demand for the return flight has been identified.

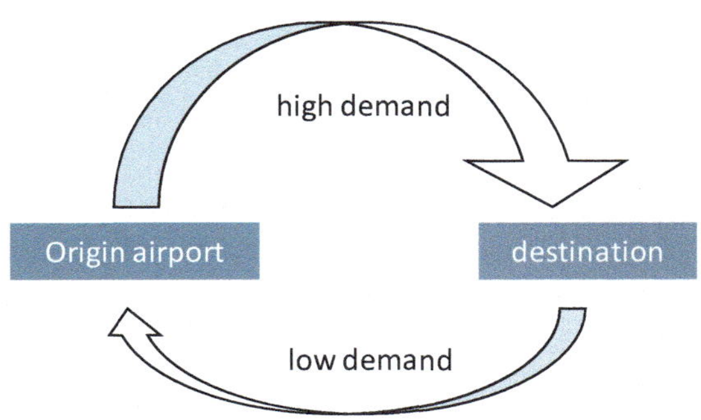


**Figure 4.3** Imbalance of demand

### 4.3.3 Volatility of demand

We started this chapter with a note that demand for air travel is derived from demand for transportation. Often, this is a luxury demand that can be postponed and this makes demand for air transport volatile to economic changes. When the economy goes up, demand for air travel tends to grow faster. When the economy goes down, demand for air travel drops faster as well. When the propensity to fly goes down due to a fallback in income or price increases of necessary elementary consumer products, the leisure traveller stays home, the EVFR traveller postpones the trip and the business traveller selects a lower booking class. In all cases, demand goes down.

Political turmoil or natural disasters can temporarily or permanently ruin demand for transport from and to a certain place. Fast economic developments can generate new demand quickly.

Mature airline aviation markets are less receptive to changes in the economy. In these markets customers can afford to fly as much as they see fit. These passengers will not dramatically change their behaviour with changes in the economy.

## Suggestions for further reading

Laurie A. Garrow: *Discrete choice modelling and air travel demand*. Routledge, 2010.

# CHAPTER 5

# Sustainability of aviation

In this Part I, on the structure of the aviation sector, we need to discuss the issue of the sustainability of aviation. Aviation is *not* sustainable in its present constellation. This has become a serious issue since the world community has adopted climate change as a prime political issue.

Aviation consumes massive amounts of fossil fuels. It is *not* that aircraft are excessive burners of fuel, given the *distances* that they cover with their payload. The fuel burn of aircraft has gone down dramatically over the years. Today's passenger aircraft burn almost the same amount of fuel, on a per passenger/kilometre basis, as a motorcar. But, as aviation provides the greatest distance to the market, it consumes vast amounts of fuel. Today, aviation contributes to about 3% of man-made $CO_2$ emissions, but this percentage is growing due to the growth of aviation as well as the shrink of other contributors to this green-house gas emission. And with $CO_2$ emissions becoming a prime concern in society worldwide, the issue of sustainability of aviation in its present form has become an overwhelmingly important issue. Therefore there are many reasons to spend a chapter on this aspect of aviation.

## 5.1 $CO_2$ emissions of aviation

In the previous chapters, we have seen that aviation has grown spectacular, resulting from decreasing production cost and increasing intrinsic demand for travel. We have also seen that profit margins are moderate. This all results in aviation being what it is today: a *quantity-driven* transport sector where *moderate profitability* requires *vast numbers* of passengers.

Production in aviation – operating aircraft – requires huge amounts of fossil fuels: kerosene, to be precise. Every single kilogram of kerosene burnt in a jet engine generates 3,148 kilograms of $CO_2$. With an anticipated kerosene consumption of some 460 million tons in 2019, the $CO_2$ production in absolute terms is a straightforward calculation, which makes aviation a significant contributor to the greenhouse effect changing the globe's climate. This aviation contribution presently stands at some 3%, but as aircraft emit their greenhouse gasses at high altitudes, the net effect is believed by scientists to be larger.

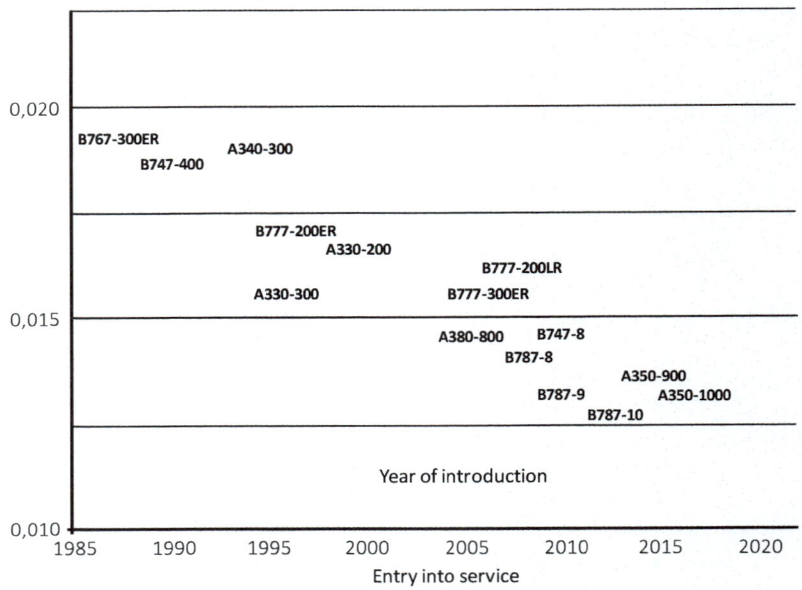


**Figure 5.1** Improved fuel efficiency

The combination of a quantity-driven sector and its massive $CO_2$ emissions cannot continue forever. Nevertheless, the industry anticipates on a further growth of aviation, expecting a doubling of 2016 output by 2035, according to the forecasts of the major suppliers.

## 5.2 The Paris Agreement

In 2015, the majority of states agreed on a treaty governing the reduction of $CO_2$ emissions. This "Paris Agreement on Climate Change" has been ratified by many states and has been in force since 2016. It stipulates that global $CO_2$ emissions may top in 2020 to decrease annually after that, down to the 1993 emissions levels by the year 2035. In 2050, the world society has to be $CO_2$ neutral, meaning that any emission should be compensated for by a reduction of the same quantity.

As said, the Paris Agreement is between states, where individual sectors like aviation are not in itself subject to the agreement. The idea is that states will, on the basis of the Paris Agreement, jointly impose reduction targets to individual sectors within the legal international frameworks of these sectors. For aviation, this is ICAO. As such, the Paris Agreement is the legal basis for ICAO to develop terms and conditions where under aviation contributes to the goals as set in the Paris Agreement. In Section 5.4, we will discuss what ICAO has developed so far, but in order to understand ICAO's endeavours in this matter, we need to discuss the specific problem of aviation with regard to the spirit of the Paris Agreement: the lack of transformation possibilities.

## 5.3 Technical solutions

The spirit of the Paris Agreement is to lead the world into a process of *transformation*, whereby the use of fossil fuels is transformed to the use of renewable sustainable energy. A very visible example of this transformation is in the automotive industry, where engine power is transforming from combustion engines to electric engines powered by battery cells. This technology can and will also be applied to the maritime sector, and to public road transport systems in general.

The specific challenge that aviation faces in this process is that transformation technologies in aviation are in an *infancy* state and *difficult to apply* on a large scale. The root cause for this observation is the simple fact that aircraft need to conquer gravity, which requires vast amounts of energy. Where surface transport modes can transform to electric traction since the size and weight of the batteries is not a limiting factor, aircraft need a sudden availability of energy to become airborne and a steady stream of energy for the cruise flight. Storing this energy in batteries is far from possible since existing batteries are far too heavy for application in an aircraft. Battery-powered electric propulsion of aircraft is in its infancy and has been limited to very small aircraft so far. Electric aircraft with payload and range capabilities useful for the aviation business is very far over the horizon, both in time and in technology development.

A possible transformation from fossil fuel to *biofuel* or *synthetic* fuel faces the problem of the vast amount of fuel that aviation consumes. Producing biofuel on such a scale would require massive amounts of agricultural areas at the expense of use for the food supply of these areas. Producing all that synthetic fuel would require massive amounts of electric energy. This is also one of the problems with a possible transformation to *hydrogen* energy, the electricity required for hydrolysis of water. As burning hydrogen results in emission of that water, doing this at high altitudes, as aircraft do, could well result in just more climate problems.

Many aviation organizations, suppliers and research institutes work hard on developing technologies aimed at reducing the $CO_2$ emissions of aviation. But we need to observe that these technologies are either in their infancy or exist in theory only. Given the timeframe of aircraft development and given the life cycle of aircraft in general, as we will discuss in Part III, we need to accept the conclusion that aviation will find difficulties in conforming to the transition process, and that aviation will certainly not be able to do this within the timeframe as set in the Paris Agreement. This observation is the basis for ICAO's initiative for sustainable aviation.

## 5.4 CORSIA and beyond

In 2016, ICAO's member states have agreed on a Carbon Offset & Reduction Scheme for Civil Aviation, CORSIA. This agreement is also known as GMBM, Global Market Based Measure. It will become legally binding to all ICAO member states by 2027. The core of this agreement is that, since aviation cannot transform like other sectors, it will, by means of a $CO_2$ surcharge, *subsidize, or even finance, the transition of other sectors*. The amount of such a surcharge has not yet been decided. As CORSIA will expire in 2035, it is apparent that it is meant as a starting point paving the way for more stringent regulations in the future. Already in its present general form, execution of this agreement will become complex.

ICAO has installed a Committee on Aviation Environmental Protection, and this committee had developed a CAEP/10 $CO_2$ standard, a standard that all new aircraft need to comply with from 2020 onward. CAEP/10 also defines operational standards on cruising altitude and descend/approach procedures aimed at reducing fuel consumption. All aircraft in production today comply to this standard. Indeed, the 10–20% fuel burn reduction of contemporary aircraft compared to the former generation of aircraft and engines is impressive. However, the problem here is that the growth rate of aviation, and with that the proliferation of the number of aircraft in operation, is much faster than the fuel burn reduction per aircraft, resulting in a net growth of $CO_2$ emissions in aviation.

On a regional scale, we observe many initiatives aimed at pricing $CO_2$ emissions. Most notably the EU has adopted an Emissions Trade Scheme. This scheme is applicable to all industrial sectors within the EU and will also become applicable to aviation within the EU from 2020 onward. The core of this system is that any organization that emits $CO_2$ needs to have allowances to do so. These allowances can be traded resulting in market forces defining the price of such allowances. The ETS is gradually getting momentum with a constant rise of the price of ETS allowances.

On a national scale, we see many countries discussing the need for an emission charge, or a straightforward tax, on aviation fuel consumed domestically. The Chicago Convention forbids the taxation of fuel for international flights, but taxation on a national scale is allowed and applied in various countries. Imposing airport taxes is also allowed and applied in many countries.

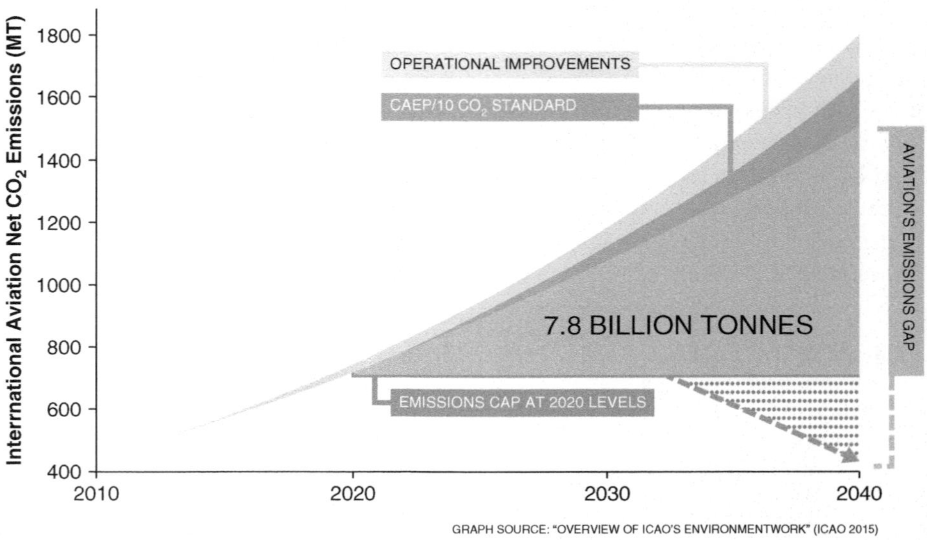


**Figure 5.2** Aviation's emissions gap

## 5.5 An inconvenient truth

From the above we can conclude that the initiatives, programmes under development and political discussions regarding $CO_2$ emissions all boil down to the same thing: an *increase* in the price of aviation products leading to an *increase in air fares*. Sooner or later, and at a rate that cannot be predicted yet, ticket prices will go up.

In Chapter 4 we saw that a main driver for the massive growth of aviation over the last decades is the constant decrease of the price for aviation, allowing more and more people to travel to the extent that air travel has become an affordable consumer commodity. An anticipated constant increase of the price for air travel could have the same effect in the opposite direction: a decrease in the demand for air travel. It is a completely unknown factor as to what extent a rise in air fares will decrease demand. In Chapter 4 we also observed that with the world becoming a global village, the perceived need for air travel has grown. An increase in the price for air travel could also lead to consumers lowering the consumption of other goods and services. An increase in the price for air travel could certainly result in passengers choosing a surface transport mode for their journey, well possible for distances up to some 1,000 kilometres.

But, uncertain as it is in time and magnitude, a constant increase in the price of air travel will ultimately influence the demand for air travel; with this observation we need to face an inconvenient truth: the quantity-driven, low-margin aviation business *cannot sustain* its present form in the face of anticipated rising air fares. The prognosis of aviation output doubling in the next 15 years therefore is very unlikely to happen without significant changes in propulsion technologies.

## Suggestions for further reading

Alice Bows, Kevin Anderson: *Paul Upham: Aviation and climate change*. Routledge, 2014.

Peter McManners: *Fly and be damned, what now for aviation and climate change*. Zed Books, 2010.

# PART II
# Airports

Airports are the places where aviation connects to ground transport. At airports, aviation becomes visible to the outsider. Here the customers of aviation gather to board a flight. Airports are the port of entry into a country, a region or a city. They connect cities and regions with each other.

Airports are complex in the sense that managing airports, managing capacities and managing processes and procedures all have specific complexities but need to be performed in a synchronized manner. Simultaneously, airport operations are vital to the operation of aviation for efficiency, cost and above all, safety and security.

At airports, many different aviation players need to cooperate. Airlines, ATC, ground handlers, MRO organizations, fuel providers and catering suppliers need to plan their operations in a coordinated manner in order to turnaround an arriving aircraft within the time frame as set by the schedule of the aircraft.

Airports are large centres of employment in aviation: most people working in aviation work at an airport. This is reason enough to devote a part of this textbook to the phenomenon of airports. We will start Chapter 6 by explaining the legal framework in which airports operate. Chapters 7 and 8 will present a macro view of the strategic level of airport management. In Chapter 9 we will discuss the services that the national state needs to provide at airports. In Chapter 10, we will take a closer look at the challenges of managing capacities at airports and the operational aspects that make capacity management of airports so challenging.

A final point crucial to airports is that they are the *bottlenecks* in aviation simply because aircraft manufacturers can produce aircraft at incredible volumes, but airports cannot grow at a comparable pace anywhere in the world.

# CHAPTER 6

# The legal framework of airports

## 6.1 ICAO regulations

Rules and regulations on airports are based on ICAO standards set in *Annexes 9 and 14*. Airports do not fly, they are where they are and stay there, fixed and firm. And they are located *within the territory* of the state. At airports, aviation connects with the rest of the world. Therefore, at airports not only aviation regulations but national laws apply.

On all aeronautical aspects – the *air side* of the airport – airports need to comply with Annex 14 of ICAO, and under FAA Part 139, the AOC, the *Airport Operating Certificate*, provides proof of this compliance. In EASA Part 139 it is called the ADR, *The Airport Design Certificate.*

Both define runway specifications and conditions, navigation equipment for approach and landing, radar coverage of the air terminal and the surface of the airport, taxi-way layout, airport approach and runway lighting, and marks and signs on the runway-taxiways. It furthermore describes how the airport needs to provide services in winter conditions, like de-icing of aircraft. It stipulates procedures on the platform.

Airports need to have an approved SMS, *Safety Management System*, in place. This is very detailed as it needs to describe in detail how the airport will react to incidents and fatal accidents within its airport area. An established fire department needs to be in place, and a chain of command. An established fire department needs to be in place, and a chain of command needs to be defined.

Thus, on all these aeronautical aspects, airports are similar everywhere in the world.

Once in the terminal, on the so-called *land side of the airport*, this is different. Here, national legislation of the country where the airport is situated prevails. This legislation is enforced by the national or local Police, or by the Military. Routine aspects of this enforcement, like checking passengers and their carry-on luggage, may well be "outsourced" to private security firms, acting on behalf, thus under the responsibility, of the enforcing state organization. ICAO's Annex 9 describes standard practices on the handling of passengers and cargo in airport terminals, but it is up to the national state to define and enforce its practices. ICAO's Annex 9 is a guideline.

Immigration of persons is governed by the state's *immigration* agency, and the importing of goods is governed by the state's *custom* agency. So, once in the terminal, one has really arrived in the country where national laws need to be obeyed.

Many airports in the world are subject to national legislation on aircraft noise and emissions. ICAO's Annex 16 provides standards on measuring units and measuring methods, and specifies aircraft noise classifications. ICAO provides standards as the basis for national rule-making. While this text was being written, the ICAO Council adopted a classification standard on emissions, subject to approval at the next ICAO Assembly.

As airport noise and emissions primarily impact the local environment at and around the airport, *local authorities* can influence the making of national legislation or impose legislation themselves, depending on how the hierarchy of authority is defined in a country. In many countries the authorities of the city, county, department or province in which the airport is situated have a huge say in its noise and emissions regulation. This can result in rules on airport opening hours, aircraft movement caps, noise abatement procedures, approach paths and surcharges on landing fees, as per ICAO-defined aircraft category.

ICAO assigns a 4-letter code to any place where civil aircraft can land and depart: an aerodrome. If one needs to find the aeronautical data of an airport, the 4-letter code of that airport will lead to such information. The first letter refers to the region where the airport is located, the second letter may refer to a specific country, and the last two letters refer to the airport itself. All aeronautical data of an airport can be found by using its ICAO code, whereas for all passenger-related elements the 3-letter IATA code prevails. Hence, for IATA, Los Angeles International Airport is LAX, and for ICAO, it is KLAX.

ICAO categorizes designated aerodromes in different categories, based on the maximum size of aircraft that such aerodrome can receive. If one looks for an aerodrome on the basis of ICAO's 4-letters, one will find the reference code of that airport, indicating what size of aircraft such aerodrome can receive. KLAX is a category 4F airport, the largest category.

| CODE ELEMENT 1 | | CODE ELEMENT 2 | | |
|---|---|---|---|---|
| Code number | Aeroplane reference field length | Code letter | Wing span | Outer main gear wheel span[a] |
| (1) | (2) | (3) | (4) | (5) |
| 1 | Less than 800 m | A | Up to but not including 15 m | Up to but not including 4.5 m |
| 2 | 800 m up to but not including 1 200 m | B | 15 m up to but not including 24 m | 4.5 m up to but not including 6 m |
| 3 | 1 200 m up to but not including 1 800 m | C | 24 m up to but not including 36 m | 6 m up to but not including 9 m |
| 4 | 1 800 m and over | D | 36 m up to but not including 52 m | 9 m up to but not including 14 m |
| | | E | 52 m up to but not including 65 m | 9 m up to but not including 14 m |
| | | F | 65 m up to but not including 80 m | 14 m up to but not including 16 m |

a. Distance between the outside edges of the main gear wheels.

**Figure 6.1** ICAO Aerodrome reference codes

Summarizing, we can conclude that ICAO's rules and regulations on airports is meant to provide maximum safety and ensure effective operations on and around airports. ICAO's regulations ensure the maximum level of international standardization. ICAO also defines the border between international air law and national law at airports.

## 6.2 Air side and land side

An airport is a place where aviation is connected to ground transportation. It is the place where the aeronautical world, with its stringent rules and regulations, processes and procedures, is linked to the non-aeronautical world. The border between the two is physically located at an airport; every airport is strictly divided into two separate parts, the *aeronautical* air side and the *non-aeronautical* land side.

Parts of the airport where the *aircraft* move and park, and *where air law prevails* are all part of air side. Parts of the airport where *passengers* move or wait and *where the national or local authority is in charge* is land side. Strictly speaking, one can say that at a modern airport with boarding bridges connecting the terminal with the aircraft, the border between air side and land side for the passenger is the threshold of the aircraft entrance door. Once inside the aircraft, the air side is where air law prevails, with the *pilot in command* or a designated representative from the airline holding jurisdiction over the aircraft.

In real life, it is more complicated as most of the passenger terminal is restricted to passengers that have gone through the security check and/or border control only. But once beyond the security check the passenger still resides at land side, be it that such passenger is *endorsed to enter air side*. ICAO stipulates that every person with the intention of entering air side, be it a passenger, a crew member or ground staff, need to be cleared to enter, and this clearance needs to be performed at land side.

For this reason, many professionals consider air side to start *behind this clearance point*. The problem with this is that the location of that clearance checkpoint can be different at airports. At some airports, the security check is performed at the gate. Most airports have this

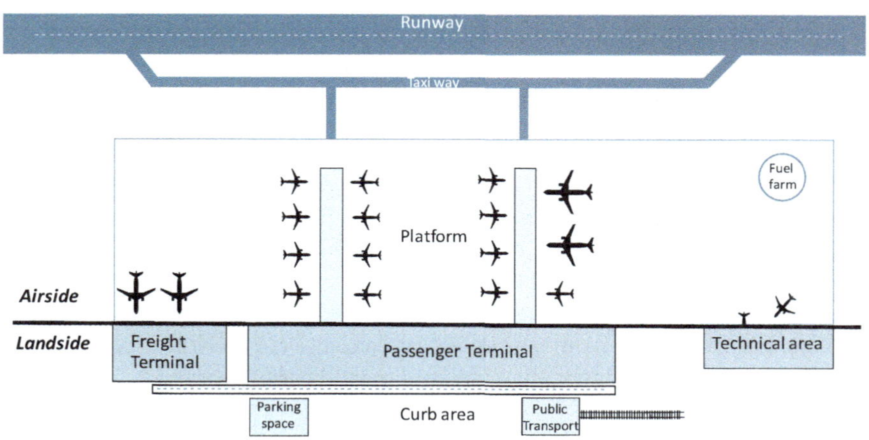


**Figure 6.2** Airport layout: air side versus land side

check just behind customs, and at some airports the first check is already at the entrance of the terminal. And at any point inside the terminal, also behind the checkpoint, the National Authority is in charge.

In the previous paragraph, we saw that there is a distinct difference in regulations between air side and land side. Air side is governed by ICAO Annex 14, a detailed document containing *binding aeronautical rules* and regulations, to be applied and enforced by the National Aviation Authority. Land side is governed by Annex 9, defining *recommendations* and common practices as a basis on which national states apply their national laws on land side. This is the reason why airports may look profoundly different at land side, inside the terminals, while at air side, the aprons, taxiways and runways, all airports look alike.

## 6.3 Nationality and jurisdiction at air side

One of the reasons why the distinction between land side and air side at airports is important is in the field of *nationality and jurisdiction of aircraft*, its crew and its passengers at a foreign airport. Here we enter the field of International Law.

Every state has exclusive jurisdiction within and above its territory. This means that a state can impose laws on any entity within its territory. This applies fully to any entity holding the nationality of the state. In principle, it also applies to entities of other nationalities, but here the sovereign state needs to *respect* entities within its territory holding another nationality. International Law stipulates in how far the exclusive jurisdiction of the state is applicable to entities within its territory holding *another* nationality. International Law defines to what extent the state needs to respect other nationalities within its borders.

This element of International Law is very important to aviation, as aircraft fly and land in territories of other states. A US aircraft registered to an airport in a European country needs to obey the laws of that European country. But that country in return needs to respect the US nationality of that aircraft, its owners, its flight crew and its passengers. When this US aircraft is parked at an airport in a European country, the laws of this country prevail. Once in-flight, when the aircraft is moving at its engine power and later flying in the airspace of the European country, US laws apply on board.

## 6.4 Rules and regulations on security

In Section 6.1 we learned about ICAO's rules for the safety, efficiency and standardization of aviation, in order to avoid *unwanted* incidents and accidents. Security deals with avoiding accidents that are *deliberately* caused by crime or terrorism. As aviation is vulnerable, international and very visible by nature, it has become a *regular target* of terrorism.

Aviation security is based on the philosophy that an aircraft should only depart when that aircraft is secure, because the possibilities to secure an aircraft in-flight are very limited. Therefore, most aviation security operations are concentrated on airports. And in order to secure the aircraft operations at the air side, airport security very much concentrates on security in the airport itself.

Rule-making on security is relatively new as this threat did not exist when ICAO was formed in 1947. Only in the 1970s, when aircraft hijacking began in the US and in the Middle East, followed by bomb attacks on aircraft in flight and in airport terminals, did national governments start to impose rules and regulations on security, individually and in cooperation, on the basis of ICAO Annex 17, which came in force in 1974. The Universal Security Audit Program (USAP) was launched in 2002 to monitor ICAO Contracting States' compliance with the standards set in Annex 17.

Where aviation authorities are the exclusive authorities for rules and regulations on safety, the responsibility for rule-making on security is wider, involving governmental departments of Internal Affairs, which, in many countries, are one of the most powerful departments. In the US, security legislation and enforcement is done not by the FAA but by the US Department for Homeland Security. And within the EU it is done not by EASA but by the European Commission's Directorate General for Internal Policies. Their rule-making is not exclusive, nor limited, to aviation. Generic security regulations also apply to maritime transport, railways and a nation's points of entry in general.

As every state is sovereign in imposing rule-making, and as security is of national interest, security regulations are national. However, everyone realizes that national regulations can only be effective in the international environment of aviation, when these national rules are at least comparable but preferably similar. But here also, as with ICAO ruling, the national states have the right to impose security rules that are more stringent than what ICAO defines. Annex 17 is a *minimum* standard.

National governments have two distinct responsibilities with regard to the security of aviation. First, they are responsible for the security of their airports, their national airlines and their national ATC system. Second, they are also responsible for the security of *all aircraft, irrespective of the nationality* of that aircraft, its airline or its passengers, departing from their airport. Annex 17 defines the responsibilities of the *Host State*, the country of departure of an aircraft. It means that the airport of *departure* needs to fulfil the security requirements of the airport of *arrival*. Here we also find the legal basis for the fact that security legislation enforcement is primarily executed at airports.

Legislation on security changes regularly, adapting to, and reacting on, new occurrences. As terrorists are creative in finding new means of attacking aviation, the authorities also need to define new legislation on preventing such attacks. Terrorist attacks have a very high public visibility, so politicians need to act swiftly on attacks in order to ensure aviation security. Such new legislation is always based on state-treaties, and are extensions to ICAO's Annex 17. As such, we have the *Tokyo Convention* of 1963, on unlawful acts committed aboard a flying aircraft; the *Hague Convention* of 1970, on the unlawful seizure of aircraft; and the *Montreal Convention* of 1973, on unlawful acts against the safety of aviation. The latest development here is the Beijing Convention of 2010 on the unlawful transportation of biological, chemical or nuclear goods. As this treaty has not yet been ratified by all ICAO members, it is not in force at the moment of writing of this book.

Especially passengers and even airlines sometimes question the effectiveness of new regulations, as they can have a severe impact on their travel experience. Sometimes security regulations can have a negative effect on safety regulations. After the 9/11 attacks in 2001 it was declared that for security reasons cockpit doors had to be armoured and could only be opened from the inside. In 2015 a depressed pilot, alone in the cockpit, deliberately crashed his aircraft, killing all on board. His fellow crew members could not stop him as the cockpit door was firmly closed from inside.

CHAPTER 7

# Types of airports

At first glance, most airports look alike, but they are not. Apart from their vast differences in size, they may differ profoundly in terms of their function within the city or region in which they are located and in value of the airlines operating within them.

There are about 40,000 airfields around the world – places designated as civil aerodrome and identified by an ICAO aerodrome code. Some 4,000 of these aerodromes are airports handling commercial aircraft, passengers and/or cargo. Of these, some 100 airports handle more than 15 million passengers per year.

What makes an airport big, or relevant for aviation operations? The basic parameter for airport function and relevance is in the location of that airport.

## 7.1 Airport location

Demand for aviation connections come in many forms. Business travellers need to fly from one major city to another; vacationers want to fly to tropical beaches, while cargo finds its destinations in industrial areas. Understanding an airport starts with understanding its location; airports are defined by the *socio-economic function* of their location.

### 7.1.1 Catchment area

The function of an airport resulting from demand for air transport to and from the location of that airport is defined by its catchment area. The catchment area itself is defined by the *surface transport time* to and from that airport, which is commonly 90 minutes; if we draw a circle around an airport location, representing a maximum of 90 minutes' travel time by ground transport to and from that airport, we have the catchment area of that airport.

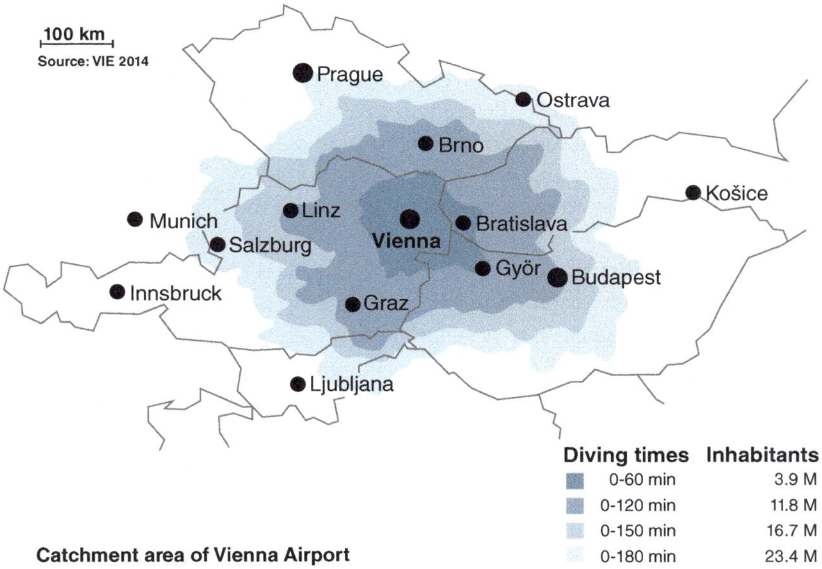


**Figure 7.1** Catchment area

Now we can make a socio-economic analysis of that catchment area, for which we can use the *DESTEP analysis*, a tool developed in social geography. DESTEP stands for Demography, Economy, Social-cultural, Technology, Environment and Political stability. By analyzing these factors in the catchment area, we can assess demand for air travel to and from the airport that is located in the centre of that catchment area.

*Demography* is quantitative, comprising the number of people living within the catchment area. It is fair to say that in general, demand for air travel will be higher in a vastly populated urban region than it is in a scarcely populated rural area. But the impact on aviation of the population size needs to be considered in close conjunction with the second element of our analysis, i.e. economy.

*Economy*, or the economic wealth of a catchment area, defines to what extent people living in that area have need for air travel and the means to afford it. As such, demand for air travel in a lightly populated Scandinavian city can be more than that in a highly populated city on the Indian subcontinent. Economy also means analyzing the number of enterprises in the catchment area which generate demand for air transport.

The third element is the *socio-cultural* function of the catchment area. Every city or region has its socio-cultural specifics influencing demand for air travel. Regions with a multicultural population might generate VFR demand for air travel, while wealthy countries at northern latitude may generate leisure demand for air travel.

*Technology* is closely related to the first three factors and indicates how far technology has proliferated into the catchment area. A population widely using smartphones and having access to internet fly more than a population not having access to advanced technology.

*Environment* is the element in the analysis that tells us more about the specifics of the airport location. For example, it may be foggy or known for its large bird populations. The airport itself can be restricted by noise regulations.

*Political stability* is also a factor defining the overall attractiveness of the catchment area, as locations with a stable political climate will overall do better than a location in political turmoil.

Catchment area relates to the attractiveness for departing passengers, but we need to observe the attractiveness of the airport also from a point of view of the *arriving* passenger. How attractive is the airport as destination? Here it is obvious that an airport with a large catchment area is by definition an attractive destination, like London, New York or Shanghai. The specifics of a leisure destination, however, lay in the imbalance; this type of destination may be popular for visitors but less important for the local population.

### 7.1.2 Surface connectivity

We saw that catchment area is defined by ground transport time to and from that airport. Ground transport time is not only a function of distance but also a function of *travel mode*. Thus, the transport modes that connect the airport with its region also contribute to the size of the catchment area. This is why airports are often linked to a country's (high-speed) rail and highway systems. Airports close to cities are often linked by the city's urban transport system. Travel time by car is not only defined by distance, but also by road capacity and traffic density.

### 7.1.3 Surface transport alternatives

For travel distances up to 1,000 kilometres, surface transport modes like cars or trains could be viable alternatives to air travel. If these alternatives do not exist – for example, on islands or in mountainous areas – air travel may be the only practical mode of transport. It is fair to say that the fewer the alternatives, the higher the demand for air travel within the catchment area. Islands often have big airports, and island states often have big airlines. In continental Europe, which is made up of a patchwork of states, we see that railroads are nationally organized, but international connectivity is underdeveloped. This is an important reason for the fast growth of aviation in continental Europe. In large countries like China we see a rapid transformation from air travel to surface travel by the expansion of China's high-speed rail system. In Japan, we see that domestic travel often needs aviation as many areas are too mountainous for rail systems.

### 7.1.4 Distance between airports

There is no general rule-of-thumb on a minimum distance between airports, but for understanding the impact of a catchment area for a specific airport, the presence of other airports within that catchment area has an impact by definition.

Many large cities have more than one airport. Here we often see all airports serving a specific function, like international versus domestic operations. In situations where cities with an airport are located close to one another we often see competition between airports, whereby the airport closest to the city, such as Haneda Tokyo, has the advantage over a competing airport further away, like Narita. But it also is a matter of airport accessibility and airport quality: The airport of San Jose CA can compete against San Francisco Int., airport for the premium passenger in Silicon Valley.

### 7.1.5 Hub location

So far, we discussed the location of an airport in relation to the city or region where that airport is located. However, airports can also have a function isolated from their region, i.e. *location on the globe* in relation to the *airway routes*. If an airport is located in the centre of airway crossings or in the center of a continent, that airport could well benefit from its hub location. Many airlines connect their flights at their home airport in order to be able to offer more connections, and they do so at airports that are suitably located for such operation. As the vast majority of air travel concentrates on the *northern hemisphere*, it is fair to say that the more north an airport is located, the better its hub potential is. Indeed, Detroit or Helsinki have a great hub location, and are great hub airports.

**Figure 7.2** The Northern Hemisphere

## 7.2 Different types of airports

Demand for air travel comes in many forms, as mentioned in Chapter 4. We therefore see differences between sorts of airports, whereby many airports in the world today are a *mix* of various airport functions. Therefore, the first distinction between airports is in its *functionality*. It is important to realize here that many airports have a *combined* functionality.

A second distinction between airports is *international* versus *domestic*. An international airport has a custom and immigration facility as well as the capability to perform security checks in concurrence with the requirements of the arrival country. In general, international airports are bigger than domestic airports and big airports are often international airports. And more and more, regional airports become international airports.

A third relevant distinction between airports is in the *presence of a home carrier* at that airport. Airports that are the home airport of one or more airlines are by definition bigger and busier than airports only used by visiting carriers. Home airports generally have a larger technical site as the home carrier often bases its technical facilities at the home airport. We will discuss this in Section 7.3.

### 7.2.1 O+D airports

Most airports are so-called O+D airports, where the O stands for *Origin* and the D for *Destination*. Airports around mega cities are predominantly O+D airports where the airport(s) fulfil the air transport demand *from and to that city*. London Heathrow, JFK or Haneda are textbook examples of O+D airports. These airports have a catchment area that justifies the capacity of that airport and the capacities deployed by the airlines operating on that airport.

Almost all large O+D airports are international airports and often a home airport of one or more carriers.

### 7.2.2 Hub airports

These airports are often bigger than one would expect on the basis of its catchment area. Airports with the right hub location are used by its home airlines to connect their flights. Examples are Detroit Metropolitan located at a high Northern latitude, or Zurich International centrally located on the European continent. The largest airport in the world in terms of passengers handled, Atlanta Hartsfield, bases this position on its hub function and not on its catchment area.

Hub airports are almost always international airports. They are by definition home airports for airlines as it is the *home carrier that actually performs the hub operation*. The airline connects its flights at the home airport. By doing so, the airline can offer a multitude of connections to its markets and generate aircraft load factors on its flights higher than would have been on O+D demand at the home airport only. In Part IV we will deal with these airline hub strategies.

### 7.2.3 Regional airports

Large O+D airports are located at the large cities, often the capital of the country. But most countries have more cities, and in larger countries these secondary cities have a demand for air travel to the capital. In countries with large cities, demand for air travel between these cities is often also large. Such secondary cities have an airport, the regional airport. In countries with large domestic markets, like the US or China, regional airports can be huge as well.

Traditionally, regional airports used to be small, operated by smaller aircraft and with connections to the capital of the country only. This has changed drastically over the last two decades, with low-cost airlines heavily operating from those smaller airports and thereby transforming the small regional airport into a full-size airport with many international connections. The continuing process of *urbanization*, whereby cities worldwide grow faster than the economy, also leads to regional airports becoming major airports. Internationalization of regional airports is further enhanced by the liberalization of airline markets, allowing foreign airlines to fly to secondary destinations in another country.

As LCCs operate with many home bases, regional airports can become home airports for a part of the LCC operation.

### 7.2.4 Leisure airports

Leisure traffic – people going on holiday – forms a sizable part of demand for air travel. Airports at typical holiday spots are often typical leisure airports. These airports are used by passengers going on holiday to that tropical island, and not so much by people residing at that island. Leisure airports are therefore different from O+D airports; leisure airports only have the D for destination. And indeed, all passengers arriving at such airports need to depart from that airport as well. Think of Palma de Mallorca Airport in the Mediterranean, or Phuket International Airport in Thailand and you understand what a leisure airport is. And indeed, they can be big.

Leisure airports need to be international, as demand for holiday flights is an international market. Leisure airports are seldom home airports. A leisure airport is typically a destination airport for the airlines; however, LCCs sometimes have a large leisure airport as an operational base.

### 7.2.5 Cargo airports

Cargo is a sizable part of demand for aviation. Cargo that is transported in the belly of passenger aircraft will arrive at the same airport as the passengers. Most larger airports therefore have a cargo operation. Much cargo however is transported by dedicated freighter aircraft, and they can bring the cargo to airports with access to cargo trains, highways and often in the middle of industrial areas close to the final destination of the cargo. Liege in Belgium and Memphis in Tennessee are typical examples of airports that find their relevance primarily in cargo.

Cargo airports are international airports, as the cargo business is international, or even intercontinental, by definition. Cargo airports can be home- or hub airport for parcel operators. These operators use a hub system, and at the larger hubs they have technical capabilities available.

### 7.2.6 Refuelling airports

Today's long-haul aircraft can fly from one continent to the other without the need for refuelling. But this has not always been the case; in the past the range of long-haul aircraft was limited, and these aircraft needed to refuel en route. Airports located on these routes became relevant for refuelling. Shannon on the West coast of Ireland, Anchorage in Alaska and Dubai in the Middle East were logical refuelling airports. Passenger aircraft today do not refuel here any longer, but as long-haul non-stop flying requires additional fuel, many dedicated freighter aircraft still use these airports to refuel. The airport of Dubai has developed a high relevance as hub airport with the emergence of its home carrier Emirates.

## 7.3 The airport/airline relationship

In Section 7.2 it was mentioned that the existence of a home airline on an airport has a great influence on the functions and the size of that airport. Here we touch upon a complex element in understanding airports. We tend to look at airports as autonomous entities: we speak about airport performance in terms of passengers and aircraft movements handled. We count the number of connections that an airport offers. However, airports do not do all this themselves; *airlines* generate passengers and produce connections. Airports and airlines need each other and *jointly* produce the value proposition to the passenger.

This is even more the case with airlines that use an airport as their *home base*. At larger international airports where home airlines reside, these home airlines generate around half of all the operations at that airport. At the home airport, the home airline can even be the dominant player, as is the case at Atlanta Hartsfield, home airport of Delta airlines.

At many of such airports the home airlines, or the alliance of which the home airline is a member, have their own terminal. Managing this terminal is often done by the airline, whereby the home airlines execute part of airport management at land side. At air side the aircraft movements of the home carriers mix with the movements of the so-called visiting airlines, airlines that use the airport as destination airport in their network.

Hub airlines use the home airport for connecting flights, and the hub operation of the home carriers can be responsible for around 50% of all aircraft movements at that home airport. The value proposition to the passenger of this connection is defined by the quality of the transfer airport. The quality of the home airport therefore is an important competition tool for the hub airline. Here we see the rationale of many hub airlines managing (and sometimes even owning) their terminal at the home airport.

Additionally, home airlines often base their technical facilities at their home airports. Large airlines tend to have large facilities in terms of area size; financial turnover; and, importantly, employment. Here we see the home airport acting as an important entity for the local economy, both for employment and in terms of socio-economic relevance.

## 7.4 The local impact of airports

With the growing importance of air travel in today's societies, the importance of airports to their surroundings is also growing. One can easily say that megacities like New York or London could not function without large airports that connect these cities with the rest of the world. Global connectivity is what an airport generates for its city or region that is

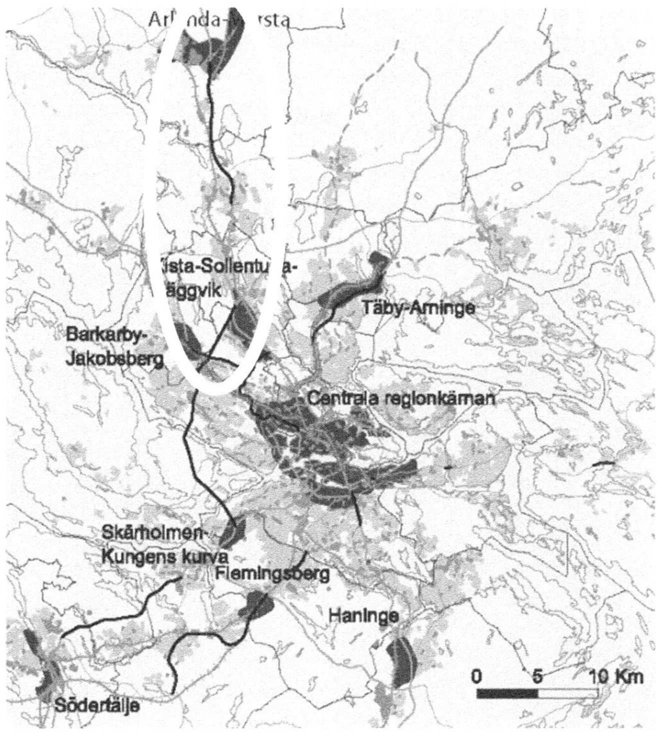


**Figure 7.3** City developing to the airport

Note: After Stockholm constructed its new airport some 60 km north of the city in the 1960s, the city developed towards the airport, with Arlanda-Stad as an important business centre.

indispensable to economic, industrial, financial and socio-economic activities. Internationally oriented businesses reside in a close vicinity to an airport offering connectivity to the world.

As aviation is labour intensive, airports generate employment, especially airports used by airlines as home airports. Most of this employment requires higher education levels; therefore airports that are home to one or more airlines generate high value employment. As aircraft operations and certainly aircraft repair and overhaul are high-tech business, airports can become a technology engine for their region or even for the nation.

Here we often see a circular development process; the existence of an airport attracts business to reside there, in itself enhancing the attractiveness of that airport as destination. This circular development is now leading to the development of *airport cities*, in which airports become a vital place unto themselves. What we also often see here is that the city or region itself develops towards the airport.

In many places in the world, tourism is a significant economic sector. In typical leisure spots, like tropical islands, it may well be the main source of income for the region. Airports provide access to these leisure spots and connect tropical islands with the rest of the world. Thus, tourism requires an airport to develop.

These crucial benefits do not come for free. Airports occupy vast territories of land mass, often located close to metropolitan areas. These vast masses cannot be used for other metropolitan functions. Areas in close vicinity to an airport are limited in the height of constructions as climbing or approaching aircraft need sufficient clearance. Access to airports require infrastructure in the form of railroads and freeways. Approach- and climb-out areas for aircraft are even larger in size, resulting in airports that have a huge impact on the environment. Dense surface traffic as well as noise and pollution of large areas of a city or region are the prices that are paid for the connectivity to the world.

CHAPTER 8

# Airport economic management

In this chapter, we will look at airport operations from an economic angle. How can we measure and rank airport performance, and what are the cost drivers for airport operations? How do airports generate income and how do airports plan their capacity towards the future?

## 8.1 Airport key performance indicators

Airport performance indicators look fairly simple at first glance but understanding the performance of airports requires a good understanding of performance indicators that contain hidden complexities.

Airports can be ranked in different ways: by the number of passengers handled, by the number of aircraft movements or by the number of connections that are offered. One can rank airports by the number of airlines visiting that airport; a new way of ranking large hub airports is by determining the relative position of one or more *airline alliances* at that airport.

If we consider an obvious KPI, i.e. *passengers handled*, we see complicating factors making this KPI one to look at with care. First, an airport counts every passenger twice: a departing passenger will likely sooner or later return as an arriving passenger at that airport. An O+D airport therefore reporting 50 million passengers actually had 25 million passengers twice at its airport. The second problem with this KPI is that airports do not discriminate between passengers. It does not measure how far the passenger started travelling from the airport, and it does not indicate whether this was a wealthy business passenger or a backpacking student. Therefore, the KPI passengers handled only provides information on the airport's *quantity* performance, not on the airport's relative value to the aviation business.

Double counting is also done with the KPI *aircraft movements* as this KPI comprises either a take-off or a landing. We call it Air Traffic Movements (ATMs). As all aircraft landing at an airport also depart from that airport again, ATMs by definition come in pairs. Hence, the airline reporting 1,000 ATMs actually had 500 visiting aircraft. Also, here, the KPI does not reveal how far the aircraft had flown or will fly as a result of its pair of movements.

More complicating is the fact that an ATM tells us *nothing* about the number of passengers arriving or departing with that movement. An ATM generated by a small turboprop aircraft is identical to an ATM of a super Jumbo.

When we realize that *air side capacity is managed by aircraft movements, and land side capacity is managed by number of passengers*, we have the main cause for the logistic puzzle at many larger airports. The two do not have a linear relation: large O+D airports like LHR or JFK receive predominantly large aircraft, so many passengers per ATM. Hub airports, on the other hand, receive many smaller aircraft that feed the hub, so less passengers per ATM. An airport ranking based on passengers handled looks different from a ranking based on ATMs.

In airport KPIs we see quantities, but we do not see the qualitative reality behind them. That is also the case with the *number of connections* that are offered from an airport. First, we do not see in which *frequency* this connection is offered. A connection offered once per week will not have the same impact for an airport than a connection that is flown every hour. Second, we do not see the *distance* to that destination, whereas a longer distance could mean a higher value of the connection: the longer the distance of a connection is, the less that connection can be produced by surface transport.

Hub airports – airports that are used as the home airport of a hub airline – offer many connections, all produced by that hub airline or its alliance partners. Here we have the problem that a feeder connection may be of the benefit to the hub carrier, without generating any economic benefit to the city or region of that airport.

At hub airports, we have another relevant KPI to consider: the *Minimum Connecting Time* (MCT). This is the time required for an airport operation to transfer a passenger and his or her luggage from an incoming flight to a departing flight. This MCT is an important competition tool among hub airlines. As passengers seek the shortest total travel time, the quality of a hub connection is defined by the transfer time. The lower the MCT, the lower the total travel time is.

Airport KPIs are important, but they all concentrate on quantities, making it difficult to rank airports on how successful they are in relation to their specific function(s). And these specific functions itself are closely related to the specifics of the location of the airport.

## 8.2 Airport economics

Airports need to operate in a cost-effective manner, and in this paragraph, we will see that this can be a true challenge.

### 8.2.1 Airport cost

Airports are capital intensive entities. The costs of constructing runways, taxiways and aprons are immense, as this infrastructure needs to be long, large and strong enough to carry heavy aircraft. The same applies to the construction of terminals, concourses and gates: high up-front investments that need to be paid back over a long period of exploitation. Airports have large *long-term liabilities* on their balance sheets. They need sufficient aircraft and passengers/freight handling to pay for the interest and amortization of heavy loans required for financing airport construction.

The *operational expenses* of an airport are also massive. The cost for maintaining the airside infrastructure, as well as the landside buildings, the energy bill for airside lighting and terminal lighting and climate, and the cost of providing and maintaining the vast digital infrastructure of the airport, all contribute to sizeable operating expenses, to be earned back by charging the users of all that infrastructure.

The main challenge for airport economics is the fact that most of the cost described here needs to be incurred, *irrespective* of the number of aircraft movements made or number of passengers handled. In the next chapter, we will see the aeronautical services an airport needs to provide under ICAO Annex 9 as well as the fact that all services need to be in place, even if an aircraft only operates at that airport once per week or even if passengers are handled only occasionally. There is not much marginality in the operating expenses of an airport. This is why many secondary airports, or seasonal airports, have difficulties in covering the costs.

### 8.2.2 Airport income

Airports are not allowed to make profits on their core activity, i.e. handling aircraft and passengers. The Chicago Convention stipulates that airports must provide their handling services against *nominal cost*. But airports are free to determine how high or low these nominal costs are.

Airports charge the airline a landing fee for every landing (and take-off). This charge is based on the aircraft's weight or seize, and often also the time of the day of that landing and the noise and pollution levels of the aircraft. Airports charge the airline also for every departing or transfer passenger. These charges form the *aeronautical income* of the airport and should cover the cost of producing the services, including the services required to cope with emergencies and fatalities.

Airport landing charges are increasingly used to regulate the operations at that airport. Airports wishing to ban noisy or polluting old aircraft can increase landing fees for such aircraft. Airports that want to avoid night operations can increase its charges for landing in the night. Fact is that the landing charge is only a very *small fraction of the total operating cost* for a flight. For deciding on aircraft-type deployment by the airline, the aircraft's landing charge is not very relevant.

Airports act like real estate companies for generating *non-aeronautical income*, where profitability is allowed. They rent out terminal capacity, including the electronic and mechanical infrastructure inside the terminal, to the airlines, floor space inside the terminal to retail and food-and-beverage companies and car-parking space to the public. At large airports they develop utility buildings, renting out office space to international companies and cargo-handling space to freight forwarders.

Airports need many passengers for generating income, and it is the number of passengers handled that is the main KPI for assessing the income potential of an airport.

Airport financial management may decide to put all aeronautical and non-aeronautical income in one basket, the *single tile* method. At many airports, these two income streams are accounted separately, the *dual-tile* method. The latter method gives airport management a clear indication of the income potential of that airport. If income is high enough, on the basis of the number of passengers handled, that airport may be able to be profitable.

### 8.2.3 Airport profitability

Probably the main reason why the academic interest in airport economics is fairly low, is because not all airports need to be profitable. For many airport owners, profitability is not the reason for exploiting the airport. For owners like national states or local governments, an airport is regarded as indispensable infrastructure for the state, the region or the city. In the previous chapter, we saw that airports have an important function in connecting the city or region with the rest of the world, whereby the airport as connector does not need to be profitable in itself. The airport here is an enabler of economic prosperity; think of an airport on a holiday island, which makes tourism possible.

But there are also instances where airports are capable of generating profits. Here we see a tendency of owners deciding to privatize the airport, selling the airport to investment entities that exploit the airport for reasons of profitability. Large airports are capable of generating profits on the non-aeronautical activities, most notably the large airports that handle many passengers. Many passengers mean income from the service fee charged to the airline on a per passenger base. With many passengers handled in the terminal, this terminal becomes an interesting spot for commercial concessions for suppliers of food and beverages and for retail. Here we see the terminal becoming a shopping mall. Airports with a large catchment area and strong origin markets can generate sizable income from parking fees. Indeed, the enormous parking garages at large airport provide an important service to the departing passenger, generating strong income for the airport. Large airports, centrally located and offering many relevant connections, can become an interesting office location for international enterprises. Here we see airports exploiting office space, a lucrative form of generating income. Airports that are the home airport of a large airline can earn from renting office space, terminal capacity or land area for technical facilities to the home airline. All in all, large hub airports located near large cities are profitable entities, often bought in full or partly by investment companies.

## 8.3 Airport ownership

There are no rules or regulations in air law on the ownership of airports. Therefore, airport ownership is regulated at the state level, but ownership varies between countries. In some countries, all airports are owned by the *state*. In many countries airports can also be owned by a *city council or municipality*. Larger airports can be *privatized* and owned by *investment companies*, e.g. Vinci in France. In France, the highway network is privatized to companies exploiting infrastructure, often with capital coming from pension funds. The latter are the investors in commercial real estate. In airport privatization, we see the emergence of airport owners that own and exploit airports both as infrastructure and as real estate. At large hub airports, notably in the US, the *home airline* can well be the owner of its terminal, sometimes even including the airside ramp at that terminal.

Fact is that it is not so important as who owns the airport, as the liberty to act is very limited for the owner. In Chapter 6 we saw that at land side, airport operations are regulated by national rule-making in accordance with ICAO annex 9, and air side is strictly regulated by ICAO annex 14. We also learned that landing rights for airlines are regulated by state treaties. Within the EU all airports are considered EU airports free for use by any EU airline, subject to any multilateral treaty signed by the EU. The chapter also noted that environmental regulations are set by (local) governments.

Airport management needs to comply with all of this, irrespective of who owns the airport. It has remarkably few powers in deciding which airlines are allowed to use the airport. In Chapter 5 we saw that landing rights are set in treaties between states, and within the EU all airports are considered EU airports free for use by any EU airline. This effectively means that, regarding international aviation, an airport management has *no decision power* on determining which airlines use the airport.

Concerning the controlled airports within the EU, airports where the number of available slots is limited, an *independent slot coordinator* assigns the available slots. So even here, airport management has no say on which airline gets what slot.

## 8.4 New airport planning

Airport planning is long-term planning in a quickly changing aviation environment. What capacity is needed ten years from now, how will aircraft look like in 20 years from now? Nobody knows, but airport planners need to do long-term planning on the basis of these uncertainties. We have seen that the place of aviation in society is constantly changing, and that the characteristics of an airport are a function of its relevance for the city of region where it is located. We also need to note that aviation is not only growing but that business models of airlines are constantly changing as well. This all means that it is extremely difficult to determine what kind of airport and what capacity is needed in the long-term in a certain city or region. Therefore, the key word in airport planning is *flexibility*.

Airport planners design flexible terminals, flexible in the handling of domestic versus international traffic, short-haul versus intercontinental flights, wide-body aircraft versus narrow bodies, business versus leisure traffic, and full-service carriers versus low-cost carriers. All categories have their specific requirements in terms of services, space and capacity. In order to remain efficient, relevant and valuable, airports need to be flexible, with room for expansion.

We also saw in Chapter 4 that demand for air transport fluctuates in many ways, with moments of peak demand versus periods with weak demand.

Airports optimized for incidental peak demand will likely not be very profitable; an oversized airport has insufficient income from passengers to cover the investments. Airports optimized for weak demand will turn into a nightmare for passengers and airlines, operating in a constant state of chaos and permanent disruption. The optimum lies somewhere in between.

Airline planners look for the moments of peak demand and decide for which demand quantity they will plan their airport. It is common to take a time between the 6th and the 30th peak hour in a defined period and plan for this. Large international airports with more constant demand may well plan on average 5% of the most busy hours of that airport.

## 8.5 Airport competition

We saw that aviation is a very competitive business; the question to what extent this is also the case for airports is a complicated one. One can say that the competitive position of an

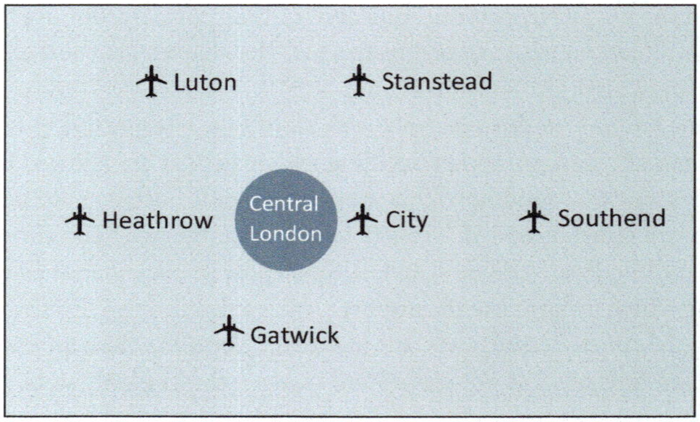


**Figure 8.1** Airports around one city

airport is not in the airport itself, but in the position of the location of that airport. Airports like Boston Logan and Malaga Costa del Sol have no competition as their positions are based on the attractiveness of the greater Boston Area and the leisure attractiveness of the Costa del Sol, respectively.

This view changes rapidly if we consider airports that are in a close vicinity to such airport. The passenger travelling to the greater Boston area could also fly to Providence, and with a final destination somewhere around Malaga, a passenger could well fly to Seville. To what extent airports compete on a regional scale depends on the quality of the surface transport modes and on the connections that airlines offer from these airports.

Hub airports can be fierce competitors. Passengers travelling from South Asia to North America mostly transfer at a European airport. All hub airlines connect their flights at the home base. Here we see customer preferences for a specific transfer airport, requiring airport quality for competitiveness. The main competition driver at hub airports is their MCT (*minimum connecting time)*. Assuming that the passenger wants a quick connection, the airport assesses on the basis of its overall logistic performance: the minimum time in which a passenger and their luggage can transfer from an incoming flight to their departing connecting flight. Hub airlines can base the connections that they sell on the basis of this MCT. Thus, the lower the MCT of the airport, the stronger its position as hub airport can be.

Many large cities have more than one airport. Here we often see distinctions in airport function, like international versus domestic or business airport versus leisure airports. This function differentiation diminishes competition between airports.

## 8.6 Airport council international

ACI is an international organization for airports. Founded in 1991 and residing in Montreal, ACI has close ties with ICAO as well as IATA (International Air Transport Association). ACI defends the interests of airports in multinational rule-making by ICAO. ACI assists airports in compliance by auditing them. ACI is an important provider of data on airports; its website is indispensable for any scholar researching airport topics.

CHAPTER 9

# Aeronautical services at the airport

Airports not only need to comply with all rules and regulations in order to operate; they also need to offer various services to their users. These services are provided by the ANSP (*Aeronautical Navigational Service Providers)*.

The ANSP is responsible for the provision of services, but at airports, in many cases, the execution of this responsibility is delegated to a specialized branch. Therefore, in this paragraph we will see many different providers of aviation services at airports, but all act on behalf of the ANSP.

The level of services that need to be provided at an airport is defined by ICAO's airport categorization. As such, airports are categorized for required air-side space, the safety at the airport and the dimensions of runways. Here, it is important to realize that airports are categorized on the basis of the dimensions of aircraft that they can receive.

The services that the ANSP provides are *not* free of charge: airlines pay for these services with a *landing charge*. This landing fee may vary according to the weight of the aircraft and the time of landing. The airport may also charge for noise and pollution. In principle, a state is free to determine the landing charge, but ICAO stipulates that if an airport uses a categorization on aircraft weight, aircraft noise and pollution, the categorizations of ICAO should be applied. Furthermore, ICAO stipulates that the landing charge should cover the *nominal* cost of the services provided; airports are not allowed to make a profit on their aeronautical services.

## 9.1 ATC at the airport

The airspace around an airport and designated to traffic to and from that airport is called the airport's control zone (CTR). This CTR, as well as the airside of that airport, is governed by the airport's ATC, referred to as Tower Control (TWR) of that airport. TWR's task is to maintain an orderly flow of traffic to, from and at the airport whilst maintaining the minimum separation standards. TWR operates under the authority of the nation's ANSP. TWR is divided in two main parts: ground control and approach control.

### 9.1.1 Ground control

The Airport's Ground Control governs all surface traffic at the airside of an airport, aircraft as well as vehicles. It is very visible to the public as every airport has a high rising control tower where Ground Control is located. With a good view over the entire airport area, Ground Control governs all aircraft and vehicle movements. Larger airports use ground radar or an SMR (Surface Movements Radar) allowing airport operations at poor visibility weather conditions, enhancing the ATC function. Thus, an airport's capability to operate in bad weather conditions very much depends on the availability of SMR.

Ground Control is also present at the airside by the Airport's Airside Control. This entity is part of the airport's organization in most countries and can act on behalf of Ground Control, with the authority of the ATC organization.

One of the tasks of Ground Control is to avoid runway incursions. This is a situation in which an aircraft, vehicle or person is on an active runway. Runway incursions are a prime cause of accidents. Ground Control therefore is, certainly at large airports with a lot of surface traffic, a serious and complicated operation.

At smaller airports, the ground control of that airport may be governed by ATC that is not physically located at that small airport. This is called RVT (*Remote and Visual Tower*), whereby the Ground Control at the main airport governs the ground traffic of a smaller airport by means of live video or SMR.

Ground Control starts when an aircraft wants to depart. Here Clearance Delivery declares an aircraft fit for departure. Clearance Delivery ensures that the departing aircraft has a filed flight plan to its destination. Ground Control governs aircraft movements: leaving the parking position, starting the engines and taxiing to the runway. When the aircraft reaches the runway, it is handed over to the ASTW's Approach Control, which we will discuss in the next section.

Ground Control governs all aircraft movements at the airport's platform and its taxiways. They plan the available capacity on the basis of the aircraft's scheduled departure times. However, these scheduled times are often disrupted by complications in the ground handling of the aircraft, complicating an optimal capacity planning by Ground Control. At various airports, therefore, a new system is adapted: CDM (*Collaborative Decision Making)*. Here, the actual assignment of the airport's airside capacity is still executed by Ground Control, but on the basis of actual information provided by the airline and the ground handler.

At airports with multiple runway systems, Ground Control assigns the active runways for take-off and landing. They will do so on the basis of prevailing winds, minimizing cross winds and noise reducing regulations. Aircraft need to start and land in headwind, and noise needs to be reduced or spread during the day.

### 9.1.2 Approach control

The airspace around the airport, with a radius between approx. 30 and 80 km, and with an altitude up to 18,000 ft, is the airport's Terminal Control Area, controlled by the TWR Approach Control. The terminal area at busy airports is divided into two parts: one for *departing* aircraft and one for *approaching* aircraft.

Departing aircraft are performing a take-off and a climb-out, both initial flight phases with specific risks. Arriving aircraft need to land safely on the designated runway, also a procedure with specific risks. Most crashes take place either at take-off, climb-out, final approach or at landing. This because aircraft fly here at low altitudes, low speed and with minimal separations all adding to a high workload for the pilots. Approach Control is therefore an important provider of safety. Executing standardized procedures at these phases is complicated by differences between aircraft in their performances at climb, descend and speed. These, in turn, stem from differences in aircraft types and in their loading efficiencies: for example, heavier aircraft do not perform as well as lighter ones in terms of climb capabilities.

A departing aircraft needs to follow a specified SID (*Standard Instrument Departure*). This is the initial climb procedure of a departing aircraft, as a predetermined flightpath along a fixed route with altitude and speed restrictions or as predetermined heading or track as provided by ATC, with or without altitude and speed restrictions. This climb-out path starts at 50 ft above the runway and ends at a predetermined altitude where the aircraft is handed over to the ATC's Area Control.

The SID is often complicated. In the initial climb phase an aircraft produces *its highest noise levels at low altitude* and often above urbanized areas. The need for noise reduction at these areas define these climb paths, meandering around populated areas. Where such a detour is not practical, a *noise abatement* procedure can be in place, meaning that the aircraft interrupts its climb by temporarily cruising horizontally at low speed and low noise over a populated area, after which it resumes its noisy climb procedure.

Upon entering the airport's air terminal, an arriving aircraft needs to follow the STAR (*Standard Terminal Arrival Route*). This is an approach path defined by navigation points, rate of descend and by speed. The STAR ends where the instrument approach begins. The TWR controller takes over from there, and when the a/c leaves the runway after landing, it is handed over to ground control. After touch-down, the arriving aircraft is handed over to Ground Control.

Terminal Control is responsible for the throughput of arriving aircraft on the basis of runway capacity. When too many aircraft enter the STAR, aircraft may be diverted to the Holding Area, where the aircraft needs to execute a precisely defined holding pattern. All aircraft in the hold follow the same horizontal pattern, so separation between aircraft in the hold needs to be ensured by vertical separation.

When, for whatever reason, the pilot aborts the landing, TWR Control needs to make a Go Around path available, defined both horizontally and vertically.

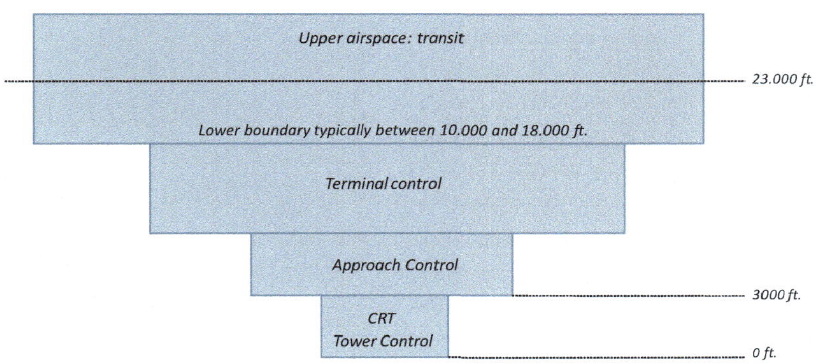


**Figure 9.1** Vertical airspace division around airports

### 9.1.3 Landing devices

The ANSP is also responsible for providing navigational devices around the airport and for providing instrument devices enabling pilots to land at the airport. Commonly used at airports is ILS (Instrumental Landing System), a radio navigational system consisting of a localizer transmitter for lateral navigation to the runway, and a glideslope transmitter for vertical guidance to the runway touch-down point, that enables the pilot to approach the threshold of the runway. ILS comes in various categories depending on the visual limitations.

There are various other ways of providing pilots with instrument guidance to the runway. They are divided into two categories: precision approach guidance and non-precision approach guidance. With precision approach guidance, automatic landings at very low visibility are possible. This may go as far as the capability to allow landings up to 75 m visibility, which is the absolute minimum a pilot needs to be able to safely taxi from the runway to the gate. To put this into perspective, in road-traffic, 75 m visibility is referred to as 'zero visibility', at which point traffic comes to a grinding halt.

The higher the sophistication of the instrument landing guidance available at an airport, the lower the potential disruption of traffic flow due to low visibility conditions. However, this comes at a cost. Maintaining the level of sophistication needed for 'blind landings' is a financial burden for the airport operator. The enormous expenses required are quite often not outweighed by the potential benefits. Therefore, not all airports are capable of the 'zero visibility' autoland.

## 9.2 Rescue and firefighting services

The most important service that an airport needs to provide on the ground is Rescue and Fire Fighting (RFFS) in case of an accident or incident at that airport or in its immediate vicinity. In Annex 14, ICAO stipulates in great detail how this should be organized and structured, based on an *airport categorization ranging from 1 to 10 defined by the dimensions of the aircraft that the airport can receive.* Most commercial airports are either category 9 or 10. Number of crash tenders, amounts of distinguishing materials, personnel required and maximum response times, all is precisely defined in Annex 14.

| Aerodrome category | Aircraft Overall length | Maximum fuselage width | Rescue and firefighting vehicles |
|---|---|---|---|
| 1 | 0 m up to but not including 9 m | 2 | 1 |
| 2 | 9 m up to but not including 12 m | 2 | 1 |
| 3 | 12 m up to but not including 18 m | 3 | 1 |
| 4 | 18 m up to but not including 24 m | 4 | 1 |
| 5 | 24 m up to but not including 28 m | 4 | 1 |
| 6 | 28 m up to but not including 39 m | 5 | 2 |
| 7 | 39 m up to but not including 49 m | 5 | 2 |
| 8 | 49 m up to but not including 61 m | 7 | 3 |
| 9 | 61 m up to but not including 76 m | 7 | 3 |
| 10 | 76 m up to but not including 90 m | 8 | 3 |

**Figure 9.2** ICAO RFFS codes

Aircraft carry large amounts of people. In case of a serious accident involving injuries, the airport needs to have medical assistance available in large quantities and at very short notice. Here the airport needs to have its logistic system in place, whereby crash tenders and ambulances can come close to the scene of the accident.

ICAO does not specify who, which entity, should organize RFFS at an airport. It may be the airport authority, a local authority or even the military. In any case, the NAA (National Aviation Authority) is finally held responsible for ensuring that RFFS are in place and functioning as per ICAO's standards. The NAA needs to license all these organizations, its staff, its equipment and its procedures. The NAA needs to audit them on a regular basis and initiate improvements as required.

## 9.3 Fuelling services

The second-most important service at an airport on the ground is the possibility to fuel an aircraft. Even if an aircraft does not land to unload or offload passengers or cargo, an aircraft may need to land at an airport to take fuel. It is safe to say that an airport can never play a role in aviation if an airport cannot provide fuel. Thus, an airport needs to be able to provide fuel, which is often easier said than done. As aircraft consume large amounts of fuel, an airport where many aircraft depart needs a steady and secure supply of a large quantity of fuel. Airports located at the sea side can rely on fuel-carrying vessels for refuelling, and airports located in land mass may need a pipeline between the refinery and the airport's fuel storage facility, often referred to as its *fuel farm*. The NAA is *mainly responsible* for providing this basic service. ICAO does not stipulate how the NAA should do that, as we see many different models here. We see airports where the NAA owns the pipeline as from the refinery, the fuel farm and the final delivery at the airport, where the airlines buy the fuel from the authority. Sometimes (a consortium of) airlines owns parts of the fuel supply chain, often in conjunction with oil companies. At most airport's it is a mix, whereby aviation authorities, often represented by the airport authority, home airlines and oil companies form complex consortia of fuel suppliers at the airport.

The actual price of the fuel at an airport is mainly driven by the cost of transportation of the fuel to that airport. National states are, by means of ICAO ruling, *not* allowed to impose taxation on fuel for internationally departing aircraft.

## 9.4 Flight information services

Every NAA needs to provide actual aeronautical information and data on its airspace and its aerodromes. This obligation is stipulated in Annex 11 of ICAO. In most countries, this obligation is performed by the country's ANSP. This information is about the use of the country's airspace, and the serviceability of navigational aids and runways. Actual information on a day-to-day basis can be released in the form of a NOTAM, a notice to airmen. FIS has two specific elements that are important to the operation of an airport: Flight Planning services and meteorological services.

### 9.4.1 Flight plan services

An aircraft is only allowed to depart from an airport when a Flight Plan is properly *filed*. A flight plan needs to be approved by the ANSPs of the state of departure, all overflying states and the state of destination of the flight. Filing a flight plan involves communication between the authorities of all these states, and it is the state of departure that needs to provide this service. A flight plan needs to be approved by a Flight Dispatcher. The ANSP can provide this service by itself, or may delegate it to the Air Traffic Management organization of the country.

A flight plan needs to include vital data on the aircraft, most notably the amount of fuel needed for reaching the destination plus the reserve fuel needed for holding, overshooting the arrival airport and flying to an alternate airport. This alternate airport needs to be specified in the flight plan, as well as eventual alternate airports en route. The aircraft's weights also need to be specified, as well as the number of occupants and the nature of the cargo. The total route, from the airport of departure to the airport of destination, needs to be specified in airways, including all points of navigation en route. Based on the aircraft's speed the actual times for overflying all locations is specified in the flight plan. This flight plan needs to be confirmed and approved by the ATC organizations of all overflying countries. After this is done, the flight plan is "filed". Aircraft can only depart under a filed flight plan.

Weather conditions are also a vital input for a flight plan. A dispatcher cannot make a flight plan without detailed information on the weather. Therefore, a meteorological service is required.

### 9.4.2 Meteorological services

Meteorology is a science outside the scope of aviation. Aviation agencies are not proper entities to perform this science. All countries have a specialized Meteorological Agency in place, a requirement for the nations' membership to the WMO (*World's Meteorological Organization)*, a UN agency like ICAO. Unification of services and specific operations for aviation is guaranteed by the fact that WMO's rules and regulations on meteorological services to aviation are identical to Annex 3 of ICAO on this subject. Therefore, in most countries the nation's *meteorological agency* is responsible for providing services to aviation on behalf of, and *under responsibility* of the NAA.

Meteorological agencies need to have a permanent communication between agencies in the world. A Flight Dispatcher needs to file a flight plan involving the airspaces of many countries. All of these countries need to provide detailed information on the actual and forecasted weather conditions during the flight and upon arrival at the destination. Fuel requirements and anticipated flying times strongly depend on weather conditions, most notably on the actual winds. Expected adverse weather conditions should be avoided by selecting alternative routing. This needs to be planned preferably prior to departure, on the basis of actual information at that moment, and on predictions.

The meteorological agency also needs to permanently measure and analyze the actual weather conditions at and around the airport, as well as forecast these weather conditions. Wind speed, wind direction, temperature and atmospheric pressure need to be monitored constantly as they are vital inputs for aircraft approach and landing procedures, and, in case of adverse weather conditions, also in departure. Atmospheric pressure at the airport of departure, the QNH, is vital information for calibrating the aircraft's altimeters. This information is called ATIS (Automatic Terminal Information System).

## 9.5 Customs and immigration

An airport can only be an *international* airport when custom and immigration are available at that airport. This state activity is executed by the state's custom and immigration agencies, so this is not a responsibility for the ANSP. However, under Annex 9 of ICAO, the NAA is responsible for *ensuring* the tasks to be performed at an international airport. Furthermore, the NAA is also designated to ensure that this activity is *performed and executed as per the standards and recommendations of Annex 9.* This annex is meant to ensure that the execution of the custom- and immigration processing should not unreasonably delay the timely arrival or departure of an aircraft, and that processing capacity is adapted to the actual need required for timely execution. Annex 9 proposes standards on custom- and immigration documentation and on control procedures.

But all is recommended and not firm as states have the final say in what prevails, border control versus ICAO compliance. At many airports in the world, custom and immigration – even when standardized – are time-consuming processes, a reflection of border control being a complex task at international airports in today's global society.

Customs and Immigration are state activities not limited to the airport, and in many countries these agencies are understaffed. For many customs and immigration agencies it is a true challenge to provide the manpower that is required at today's busy international airports.

Part of custom and immigration legislation deals with health issues, like the vaccination stamps in a passport. International airports therefore need to have a national health agency involved at the airport as well. Especially in case of an epidemic outbreak this health agency has far reaching authority on border control at an international airport.

## 9.6 Security at the airport

Implementing and enforcing security regulations is a state activity executed by the police or the military. Here, ICAO assigns the ANSP the task of ensuring that security at airports is performed as per the Annex 17 standards. Regarding security, cooperation between states is necessary. Regarding security all states need to cooperate to the extent that they execute each other's rulings. The security requirements of overflying states and the *arrival state* define the security procedures that are required at the *departure airport.* Here we see a strong stimulus for standardization as per Annex 17. The ANSP here is an important enabler between ICAO and the national security legislators.

Security operations at the airport is separated into two parts: landside and airside.

It is not an easy task to secure the *airside* of an airport. It needs control at the many entrance gates by physical control of all individuals and vehicles that enter or leave the airside of the airport. Main instrument is the *airside pass,* issued to individuals by the security agency. Enforcement of security rules is executed by the airport's *airside control*. This is a function within the airport organization, acting on behalf of, and authorized by, the national aviation authority as well as by the security authority.

*Landside* security, in and around the terminal, concentrates on scanning every individual passenger and their luggage for weapons and explosives, and on surveillance of passenger behaviour in and around the terminal. The first task is very visible and present to the passenger, and is executed by means of sophisticated scanning equipment. The second task is invisible to the passenger, and is performed by safety personnel, by using sophisticated video equipment and software. The national aviation agency is responsible for all procedures and scanning equipment being in accordance with ICAO ruling.

It is important to understand that aviation security has two different objectives: making aviation operations secure and securing national safety *against the potential threat from aviation*. The latter became urgent after the 9/11 attacks in the US in the early 2000s. In the 9/11 attacks, aviation was not the target for the terrorists but the weapon. When such a situation happens, it is mostly with aircraft in flight, where Air Traffic Control is in charge.

CHAPTER 10

# Airport capacity management

Managing airport capacity has become of paramount importance in commercial aviation. As aviation grows fast, both in number of passengers and in number of aircraft, the available airport capacity more and more becomes the bottleneck in the aviation business. Building new airports or even expanding existing airports requires time; finances; and above all, political consent between all stakeholders. In developed and urbanized countries, suitable locations for a new airport are scarce.

The second reason for airport capacity being a bottleneck is in the notion that most passengers want to depart or arrive around the same times in a day. Airports face peak hour demand; airlines all want to arrive and depart around the same time, which is based on the time preferred by the passenger. Airports need to cater to this peak hour demand, meaning that there is capacity surplus at other moments of the day.

All this implies that the available airport capacity needs to be utilized *as effective as possible*. The main problem here is that at an operational level, an airport is a mix of different logistic chains, all with their own logic and throughput limits.

In this chapter, we will take a closer look at the different capacity drivers at an airport, in order to understand the complexity of airport capacity management; synchronizing the separate logistic channels into an efficiently used airport. The chapter is divided in two parts: airside capacity, defined by the number of aircraft movements that the airport can handle, versus landside capacity, the number of passengers that that airport can handle. There is no relation between the two; an aircraft movement can be done by a 30-seat propeller aircraft and by a 600-seat A380.

Once parked at the aircraft stand, we see airside- and landside processes influencing each other; passenger handling and luggage handling starts at land side and are completed at air side. Airport capacity management indeed is a complex matter. This is why day-to-day operations at most airports are centralized at the Airport's Operations Control Centre.

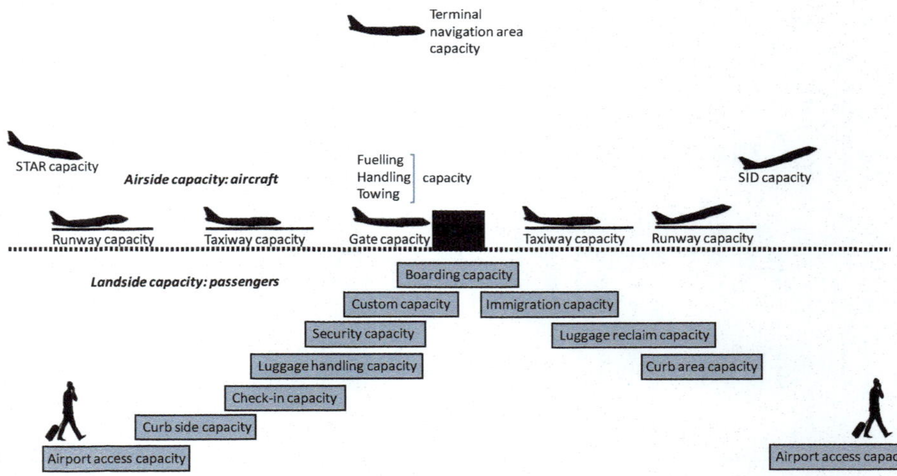


**Figure 10.1** Airport capacities

Capacity management at airports often contain a hidden paradox: in order to increase the throughput, systems should ensure a fast handling of a passenger. In order to maximize the non-aeronautical revenues, airports try to increase the amount of time spent at that airport by the passenger. This so-called dwell-time is period where the passenger spends money with the concession holders in the terminal.

## 10.1 Airside capacity: aircraft movements

Airside capacity is about aircraft using the airspace around the airport for departing or arriving, using the airport's runway for take-off and landing, and using the taxiways for taxiing to and from the apron. It is about the number of aircraft arriving and departing, irrespective of the size of that aircraft. We call this the number of ATMs (*Air Traffic Movements*).

The maximum *size of aircraft* that an airport is allowed to handle is stipulated by the airport's classification by ICAO, as discussed in Chapter 5 of this book. Here we saw that the airport's capabilities to cope with accidents are the parameter, defined by the number of crash tenders and its medical capacity.

The number of ATMs is also defined, or limited, by the airport's operating hours. Many airports are operational 24 hours per day, but certainly in urbanized areas (where the largest airports are) some are closed at night.

### 10.1.1 Terminal control area

The Terminal Control Area (TCA, or TMA: Terminal Maneuvering Area) is the connection between the airport and the 'airways'. It is the space in which arriving aircraft are continuously being put in sequence for landing. The better this sequencing is done, the more aircraft can land on a runway until the maximum landing capacity of the runway is reached due to the requirement that no more than one aircraft may occupy a runway in use for landing. It is also the area where departing aircraft climb and 'fan-out' from the take-off runway in the direction of their intended flight path.

Departing aircraft need to follow a climb-out path: the SID (*Standard Instrument Departure*). This can come in conjunction with a noise abatement procedure.

Many airports have noise restrictions. As aircraft produce most noise in take-off and climb, these noise restrictions have impact on the (allowed) capacity of the airport's SIDs.

Arriving aircraft need to follow a STAR (*Standard Approach Route*) with a five nautical-mile separation between aircraft during their final approach of the threshold of the runway. Challenge here is that different aircraft may have different optimal flying speeds and that even the same type of aircraft will have different optimum speeds depending on weight. This can become a problem on final approach where speed differences of 30 kts or more are quite common. Use of holding patterns or extensive vectoring (forcing to fly aircraft more track miles than necessary) are common solutions to deal with these speed differences. It can result in the actual throughput of the STAR falling below its theoretical capacity. This can result in aircraft being put on hold. Every STAR has a holding pattern. Being put on hold is daily practice at major airports around peak hour operation.

## 10.1.2 Runway layout

It is about how many aircraft can take off or land from a runway. Take-off capacity is determined by runway occupancy time. It takes roughly one minute per aircraft to get airborne. It is not uncommon to see aircraft depart from a certain runway every 90 seconds, sometimes less. This means that there is a physical limitation of around 40 to 45 departures per hour in the ideal situation. It assumes a runway that is solely used for take-off, and that is not obstructed by *an inter junction* with another runway of taxi-way. As many runways are faced with either or both limitations, the 45 take-offs per hour are often not possible.

Another complication with runway take-off capacity is in the *wake turbulence* generated by the aircraft's wing at and around the runway. Large aircraft generate this wake turbulence to the amount that a small aircraft cannot depart safely immediately after a large aircraft. A two-minute separation between aircraft is put into effect when wake turbulence is a factor. This separation increases rapidly when aircraft are in the A380 size category or even bigger, thereby reducing take-off capacity of a runway from 40 a/c per hour to 30 or less.

| Following Aircraft | behind | Leading aircraft | Separation minima |
|---|---|---|---|
| Heavy | | Heavy | 4 NM |
| Medium | | Heavy | 5 NM |
| Light | | Heavy | 6 NM |
| Light | | Medium | 5 NM |
| Heavy | | Super | 6 NM |
| Medium | | Super | 7 NM |
| Light | | Super | 8 NM |

**Figure 10.2** Wake turbulence separation minima

An airport's aircraft movement capacity is therefore defined by the number of runways and taxiways at that airport, and, more precisely, by the time for which the runways can be deployed on the basis of wind direction and/or noise regulations.

Also, the layout of taxiways plays an important role in the operational capacity of an airport. When taxiways are double, with one direction for each, aircraft taxiing in opposite directions can pass each other. Taxi-way layout is also a crucial factor when visibility is poor, and aircraft perform instrument landings. A runway has a so-called protected area, the area around the runway that needs to be empty in order to avoid ILS radio signal interference.

### 10.1.3 Runway dimensions

Large aircraft carrying many passengers and departing to long-haul destinations, with a large amount of fuel on board, need longer runways than short-haul, narrow-body aircraft. Therefore, the maximum size of the aircraft that can operate to and from an airport is primarily defined by the dimensions of the runway. Relevant here is that, as a rule of thumb, an airport needs a runway of some 2,200 metres to allow narrow-body aircraft operations, while long-haul, wide-body aircraft need some 3,000 metres of runway.

Let us start with runway length. An aircraft needs, by ICAO Annex 6 ruling, to be able either to reject its take-off run, or to continue its take-off with one engine failing at take-off. Various runway length units need to be defined for calculating the *allowable take-off weight* of an aircraft on that runway. The various runway length units are expressed in metres, feet or both.

The first unit is the TORA (*Take-off Run Available*): the length of the available runway. An aircraft needs to be 35 feet high at the end of the runway. Many runways have at one or both ends a *clearway*, meant to extend the point where the aircraft need to be at 35 ft. This is not part of the runway but as it is within the airport territory and is unobstructed, the aircraft is allowed to use it in its take-off procedure.

Adding this clearway to the runway length results in the TODA (*Take-off Distance Available*).

A runway can have at one or both end an extension that is not part of the declared runway, but can bear the weight of an aircraft. This is called a *stop way*, and this stop way may be added to the length of the runway for calculating the maximum speed at which the take-off can be rejected. The TORA plus the available stop way is the ASDA (*Accelerate Stop Distance Available*).

The allowable take-off weight of an aircraft is may be restricted by the TORA, TODA or ASDA. The lowest figure determines the maximum allowable take-off weight. If there are obstacles in the climb-out area of the runway (trees, buildings, antennas, hills, mountains) the allowable take-off weight may be further reduced by the requirement that the aircraft should be able to clear these obstacles with a certain safety margin, again under the assumption that one engine has failed.

Stop ways and clearway are not allowed to be assumed for landing an aircraft on a runway. For this, we have the LDA (*Landing Distance Available*) which is normally equal to the TORA or less if the runway has a *displaced threshold*.

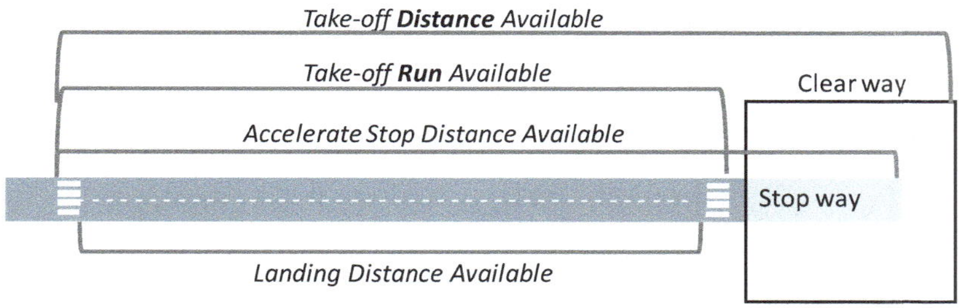


**Figure 10.3** Runway distances

Regarding the width of a runway, 45 m is the standard, but 60 m is the norm for newly constructed runways. Actual width determines the maximum allowable size of aircraft that may use that particular runway. Large aircraft like the Airbus A380 need 60 m wide runways. Most aircraft types can operate on 45 m runways. Runways with a width less than 45 m severely restrict the maximum allowable size of aircraft.

### 10.1.4 Apron capacity

After landing, an aircraft needs to taxi to its designated *aircraft stand*, a part of the airport's apron where the *Turn Around* Process (TAP) of that aircraft will be executed.

This aircraft stand can include an aircraft connected to an *air bridge*, allowing the passengers to board or disembark directly from the aircraft to the terminal. If aircraft stands are not connected to the terminal, this boarding process takes place via *air stairs and busses*. An airport's apron capacity is defined by the number of aircraft stands, and by the number of air bridges, *gates*. For the TAP of a wide-body aircraft an aircraft stand with air bridges is preferred, but at many airports this gate capacity is insufficient for handling all aircraft, which is why, at many airports, air stairs and busses are needed as well. Thus, an airport's apron capacity is primarily defined by the *number* of aircraft stands.

An airport's apron capacity is further defined by the amount of *time* required for the TAP of the aircraft at that stand. The TAP involves various processes: disembarking and boarding of the *passengers*, cleaning the *cabin*, refreshing the *catering* in the aircraft's galleys, unloading and loading *luggage and cargo*, servicing the aircraft's water and waste, and *fuelling* the aircraft for its next flight. The aircraft also needs a physical inspection to be declared as fit for flying. The TAP is terminated by the push-back procedure of the aircraft, whereby this aircraft leaves its aircraft stand, making room for the next aircraft.

Obviously, this takes longer for large aircraft making long-haul flights than for smaller, narrow-body aircraft used for shorter flights. For a long-haul, wide-body TAP a minimum of 90 minutes is assumed. For a narrow-body, short-haul operation, a TAP of 45–60 minutes is the standard at many airlines, and the Low-Cost Carriers often need less than that: within 20 to 30 minutes of arrival the LCC narrow body is ready for pushback.

In long-haul, wide-body operations, the time spent at the airport can be much longer than the minimum. The departure time of a long-haul flight is defined by its preferred local *arrival time at the destination airport.* At large international airports we see these aircraft towed away from the gate to a designated parking buffer after the deboarding and unloading processes, being towed back to a gate position some 90 minutes before the scheduled departure of that aircraft. This increases the need for apron parking space at that airport, as well as tow truck capacity.

Large wide-body aircraft need a larger aircraft stand and larger gate capacity than narrow-body aircraft. Traditionally an airport has both wide-body and narrow-body aircraft stands. The trend now is towards Stands that can receive both aircraft categories; the latest development is the so-called MARS-gate (*Multiple Aircraft Ramp System*), where either one wide body or two narrow bodies can dock. Rationale behind this is the trend for airlines to deploy wide-body aircraft at slot restricted airports. At the airport's peak hour, demand for wide-body gate capacity is higher than it is later in the day. Flexibility in gate capacity can cope with this demand fluctuation.

The TAP processes at the apron and in the terminal are performed by the airline or by ground handlers contracted by the airline. As ground handling is primarily an airline activity, we will deal with these ground handlers in Part IV of this textbook.

### 10.1.5 Fuelling capacity

An aircraft may need to fuel at its TAP, as it cannot fly to its destination without enough fuel on board. Airports need fuel supply in large quantities. At airports located close to a sea port, or airports with pipe-line connections with a refinery, this supply is normally guaranteed. At remote airports, at high altitudes or in countries with underdeveloped infrastructure, this supply of fuel can be limited, and so limiting the capacity of that airport.

Fuel at the airport is stored at the so-called *fuel farm*, from where it is distributed to the aircraft. This can take place by means of a pipeline system under the apron's surface, where *hydrant wells* are placed at the aircraft stand. At many airports and at parking places here, these wells are not available; fuel is distributed by means of *fuel trucks*. Fuelling capacity of airports is defined by the number of aircraft stands and the number of fuel trucks available at that airport. When an airport's wide-body stands are limited, forcing long-haul aircraft to fuel by means of trucks, these wide bodies, which require much more fuel than smaller aircraft, can have a significant effect on an airport's fuelling capacity.

### 10.1.6 Towing capacity

Aircraft do not taxi backwards as they are often parked with their noses towards the terminal, requiring a pushback procedure at the end of their TAP. This is performed by a towing vehicle. Here, large aircraft need a large towing truck, while smaller aircraft can be handled by a smaller towing vehicle.

Traditionally, the towing truck was connected to the aircraft's nose-wheel strut by means of a tow bar. This is unpractical as this tow bar is specific for an aircraft type, limiting the flexibility of the airline to operate to an airport. Today's towing vehicles are universal; by lifting the nose wheel a towing vehicle can push back every aircraft type or size.

### 10.1.7 Slot assignment

Airports can be unrestricted, coordinated or even declared full, defined by the number of ATMs. At coordinated and full airports within the EU, an independent *slot coordinator* needs to be in place for assigning these ATMs. Here we call them *slots*. A slot is a permission to visit the airport in a defined time frame. Airlines that operate at this restricted airport have *grandfather rights*, meaning that a slot assigned to the airline will remain in the possession of that airline under the condition that the slot is used by the airline. Airlines are even allowed to sell slots at LHR and LGW airports to other airlines, the so-called *secondary slot trading*. An assigned slot that is not used for a year can be taken away from the airline.

At important destination airports that are restricted or full, slots at that airport become valuable. Here we see airlines buying other airlines primarily for its slots.

### 10.1.8 Weather conditions

So far, we have considered the airport's infrastructure, but the extent to which this infrastructure can be used depends on the weather conditions at and around the airport.

Heavy winds with severe gust locks will limit the landing and take-off capability at that moment, and not much can be done about that. Runway capacity for landing is dependent on the probability of a successful landing. If the probability decreases, one has to cater for an increase in go-arounds. This complicates the regulation of an orderly traffic flow.

Regarding heavy fog, which limits the visibility at the airport, the impact depends on the capability of the airport's landing devices, like ILS (Instrument Landing system), and the airport's ground radar capability. A runway's landing capacity decreases if there is traffic (cars, trucks, aircraft) moving in the ILS protected area. This explains the dramatic reduction in runway capacity in fog.

Heavy rains can impact ground operations, including the aircraft's TAT as well. Snowfall leads to an airport's becoming inoperative almost by definition. One has to wait for the snowfall to stop before snow shovel deployment becomes practical.

A special type of weather-dependent operation is in de-icing aircraft prior to take-off. This is necessary for aircraft operations in freezing conditions, whereby snow, water or moist quickly become ice on the aircraft's wing surface, if not at the airport than at low altitudes shortly after take-off. Departing with ice on the wing's surface is dangerous and often results in a fatal incident; de-icing therefore is directly related to flight safety.

De-icing is performed by dedicated de-icing vehicles, spraying a glycol-based fluid over the aircraft. If done at the aircraft stand this glycol will pollute the stand, reason why at many airports this de-icing is done at a dedicated apron area. Ideally this area is close to the start of the runway. Aircraft need to take-off as quickly as possible after the de-icing procedure since the effectiveness of the glycol is limited in time.

An airport's capacity is influenced by the ways and means that the airport has at its disposal to cope with adverse weather conditions. One can say that most airports are optimized for their prevailing weather conditions. Thus, airports at windy locations have runways in various directions, and airports at high latitudes have de-icing capacity in large quantities.

Airports can handle the adverse conditions that they are used to, and are in trouble at the rare occasion when an unlikely adverse condition happens.

## 10.2 Landside capacity: passenger movements

Landside capacity is about the throughput of passengers. From the arrival at the airport up to the moment of boarding, and vice-versa, we need to realize that passenger flows come in two directions: *departing* passengers and *arriving* passengers. At hub airports we even have a third flow: the *transfer* passengers. Capacities of the various processes need to be aligned in order to avoid bottlenecks in the system.

In order to execute all passenger processes inside the terminal, IATA has defined LOS (*Level of Service*) standards in terms of the terminal floor space required for each process. These LOS figures are a guideline; poor airports may perform well below these figures, while high-quality airports may well plan their operation, with larger floor spaces per passenger.

### 10.2.1 Airport access

Landside capacity starts with the capacity of the (public) transport system(s) providing access to the airport. Considering passengers arriving by car, it is the capacity of the road/highway system to the airport. Regarding passengers arriving at the airport with their car, parking space capacity is important. To cater to passengers brought by car by others, kiss-and-ride capacity needs to also be in place.

Here we see that capacity requirements depend on the type of passenger: for departing business travellers parking space is paramount, while for the arriving business passenger taxi capacity or rental car capacity is required. For departing VFR passengers, kiss-and-ride is important; arriving VFR passengers will likely be picked up by the family.

In metropolitan areas, passengers primarily arrive and depart by public transport. This means that the airport needs sufficient railway station capacity. The challenge here is that an airport station often needs to accommodate various forms of rail transport, ranging from the metropolitan rail system to the high-speed rail system.

Airport access is a subject that is often underestimated as it is not that appealing. Airport access however is key in airport competition and airport attractiveness. Departing leisure travellers will choose the airport with the lowest parking charges, while busy business travellers will choose the airport with the shortest travel time to the city.

Once at the airport, the passenger arrives at the so-called *curb side* of the airport – the area where all ground transport arrives and where the passenger gets access to the terminal.

### 10.2.2 Check-in

The first process in the terminal is the check-in process, together with the checked luggage drop off. Traditionally this process was executed manually at the check-in desk. We increasingly see the online check-in, done by the passenger even before coming to the airport, lowering the need for terminal floor capacity and for staff.

However, not all passengers can check in on-line. For destinations where a valid passport and visa are required, the airline needs to check the validity of these documents at the departure airport. Passengers with checked luggage also need to drop off this luggage, and that luggage needs a proper luggage tag. For this reason, the industry has adopted luggage drop-off devices with limited success, as proper tagging remains difficult for many passengers.

At hub airports, check-in capacity requirements for transit passengers depend on whether the passenger can check-in at the *airport of departure*. This requires sophisticated software by the airline. In case of a transfer to another airline at the hub airport, the software systems of the airlines involved needs to be connected. The passenger receives at the departure airport its boarding pass for the connecting flight, normally with a seat assignment but often without a gate reference for the connecting flight at the hub airport. In Part V of this book we will discuss the airline alliances; indeed, check-in software integration is a key element in these alliances.

Traditionally, the result of the check-in procedure was a paper boarding pass. Today, however, airlines and passengers make use of smartphones containing a QED code with all the data required on a boarding pass.

### 10.2.3 Security check

Passengers need to be checked and secured at land side before entering air side. Thus at every airport we see the security check area. The location of this can vary: at most airports, the security check is located just after the check-in and after customs, but at some airports the security check is located at the gate or even at the entrance of the terminal.

In any case, the security check consists of a *verification* of the passenger's identity and boarding pass, a *body check* for ensuring that the passenger has no harmful items at or in its body, and a thorough check of the passenger's *carry-on luggage*. These checks are primarily performed by scanning devices and physical visitation. Throughput capacity depends on the number of lanes, the sophistication of the scanning devices, the number of staff and the content of the carry-on luggage. Regarding the scanning devices, we see a big capacity increase from X-ray to CT scanning, as the latter does not require the passenger to open its carry-on luggage any longer.

Congestion at the security check lanes increases when the airline starts charging the passenger for checked luggage. The passenger tries to take all luggage as carry-on, avoiding the luggage charge on its ticket fare, but resulting in all luggage to be checked as carry-on luggage.

The security check is the responsibility of the national land side authority. This authority can perform these checks themselves, but in many countries, we see commercial security companies executing the check on behalf of the national authority.

In principle, the security check is only required for departing passengers. At hub airports, however, we may need a security check for transit passengers that have started their journey in a country where the security procedures are considered too lenient by the country to which the passenger is heading; at most US hub airports, all international passengers need to go through security upon arrival as well.

Regarding coverage of the cost for security, we see two forms. In many countries, this check is financed by the national state as the requirement for security is imposed by the state. But in other countries the security is paid for by the passenger in the form of a security charge in the ticket fare.

### 10.2.4 Customs and immigration

Customs deals with the import and export of goods. Regarding passengers, customs checks the luggage of arriving passengers on the basis of existing import regulations on import tax duties, and on import restrictions on certain goods. Immigration deals with the passenger's identity and travel documents, such as their passport and visa. Every international airport has an immigration check upon arrival, while many countries perform such check upon departure as well. At hub airports, the custom- and immigration check is required for passengers arriving from an international flight and connecting to a domestic flight, as custom and immigration is performed at the first airport of entry into the destination country.

At many airports, custom and immigration is a sincere bottleneck. Root cause for this is that many players at the airport are commercial entities that invest in capacity with the aim of increasing profits with increasing volumes. For the national state – responsible for customs and immigration – capacity increase is a cost only, and the capacity requirements at the airports needs to be balanced against the capacity requirements at other ports of entry into the country. In holiday periods certainly, where demand for custom and immigration capacity at airports is high, national custom and immigration departments of the state often have difficulties of deploying sufficient manpower to cope with this peak demand.

### 10.2.5 Gate planning

A gate is the access point for the passenger for boarding an aircraft. Many airports use different gates for wide-body aircraft and narrow-body aircraft. At the gate the turnaround process is performed. As much can go wrong here, aircraft can be delayed, occupying a gate that is assigned to the next aircraft. All processes in the terminal are attached to a gate. It is fair to say that apron capacity and the influence of the TAP process – as described in Section 10.1.4 – find their land side inverse at gate planning. This all means that however robust the longer-term gate planning of an airport is, gate planning on a daily or even hourly basis may be a challenge. With the fast increase in ultra-long haul flights comes the unpredictability of the exact time of arrival of such flights, making tactical gate planning at international airports complicated.

Another complicating element in gate planning is that a gate both comprises of an air-side element, the aircraft stand, and a land-side element, the gate's waiting lounge. Capacity of both elements should match. With increasing seating capacities of narrow-body aircraft, the land-side part of the gate becomes too small for these high-capacity narrow-body aircraft.

### 10.2.6 Luggage handling

Passengers may come with luggage to be checked in. This luggage will be carried in the belly of the aircraft. At airports, the normal procedure is that the passenger leaves its luggage with the airline/handler at check in. The luggage will be labelled (the luggage tag) after which it is transported to the luggage handling department, often located in the cellar of the terminal building.

Here, the luggage is first screened for security purposes and then sorted for the flight number on the tag. If this flight is international, the luggage will be screened for customs.

The screening capacity is the first element to consider in capacity management at luggage handling.

At the collection point for a flight number, luggage will be stored in containers for a flight operated by a wide-body aircraft, or will be placed on luggage cart for a narrow-body flight, as the belly of a narrow-body aircraft is stowed per piece upon loading the belly.

Here also the luggage is sorted based on end destination rather than transit at destination. Many airlines also sort per booking class.

Now the luggage need to be transported to the aircraft stand, and loaded according to the sorting categories. This pre-sorting ensures a quick handling of the luggage at the destination airport. Transport capacity and travel time to and from the gates define the capacity of this process.

Upon arrival of the aircraft, luggage unloading is one of the first air-side processes, whereby transit luggage has priority. This luggage is transported to the collecting point of the connecting flight number. Dependent upon origin and final destination, this luggage may be eligible for custom and/or security at the transfer airport as well.

Luggage arriving at the final destination, or arriving at the first airport of entry into a country, are sent to the luggage belt in the arrival hall or transit terminal.

## 10.3 The cargo terminal

Aviation transports passengers as well as cargo, so airports may want to enable the players in the cargo business to operate to and from them as well. In Section 7.2.5, we saw that some airports are predominantly or even fully dedicated to cargo operations. Where for the handling of passengers a passenger terminal is a basic requirement, cargo operations require a cargo terminal at the airport. Air side operations do not distinguish between passenger- versus cargo aircraft, so no specific air-side elements are required for cargo, albeit that most cargo terminals also have aircraft stands available for full-freighter aircraft. Cargo that is transported in the belly of passenger aircraft needs to be transported from the cargo terminal to the gate where the passenger aircraft stands. This is in itself often a true challenge for airport planners. In order to lower processing times, the cargo terminal should be located close to the passenger terminal, but for the handling of dangerous goods, the cargo terminal should be located a safe distance from the passenger terminal.

In Chapter 4 we saw that cargo comes in many different forms, and an airport can only handle cargo for which the necessary infrastructure is available. International cargo handling requires *customs* to be available. Processing of perishables needs the availability of temperature controlled and *cooled storage spaces* inside the cargo terminal. This also applies to the transportation of life stock to or from an airport, requiring all sorts of infrastructure at the cargo terminal.

For every sort of cargo, dedicated elements are required at the cargo terminal in order to enable the airlines operating cargo to, from or via that airport. Handling cargo requires sufficient storage space in the cargo terminal, reason why cargo terminals can be big in floor space. As space at the airport is limited at most airports, finding and assigning space to the large cargo terminal can be economically challenging.

Inside the cargo terminal, departing cargo is prepared for flying. This means that the cargo is placed in containers or on pallets. These need to be built-up in accordance with the weight balance of the aircraft where the cargo is meant to be transferred. Building up cargo pallets to be transported in the belly of passenger aircraft can only be done when the flight plan for the flight has been generated. Thus, pallet build-up and transportation to the gate is a process under constant time pressure.

Regarding land side access most airports seek for a split between passenger movements to and from the passenger terminal versus cargo movements to and from the cargo terminal. Main device for transporting cargo on the ground is the lorry, and a constant flow of lorry's to and from the airport should not intervene with the passenger flow to and from that airport. The surface access itself is a prime driver for airports being suitable for cargo operations. It requires a central location of the airport, in conjunction with good highway infrastructure enabling the freight forwarder to deliver the cargo to its final destination quickly.

## 10.4 Collaborative decision making

The previous sections make clear that airport operations involve many different entities, sometimes with conflicting interests. At air side, the ATC organization, the airline, the ground handler and the fuel provider need to operate in a concerted manner. As all these entities have their own operations, data systems and procedures, acting in concert is a challenge in day-to-day airport operations.

At a growing number of airports, this operation is governed by A-CDM (Airport Collaborative Decision Making). In this setup, the airport operation is governed by the APOC (Airport Operations Control Centre). Airports operate on the basis of an operating plan, but as aviation operations are vulnerable to disruptions, day-to-day operations may well divert from the base plan. Aircraft may arrive beyond the STA (Scheduled Time of Arrival) on an individual base, but they may also be delayed collectively due to weather constraints at the airport. The TAP of an aircraft may be disrupted, causing an aircraft to depart beyond its STD (Scheduled Time of Departure). In order to utilize the airport's overall capacity, various entities jointly manage these disruptions at the APOC. A-CDM involves sophisticated software, either via integrated systems or, as often is the case, by connecting the data systems of the various entities.

## 10.5 Modelling and simulating

Overlooking this chapter, we see many profoundly different processes with different throughput capacities and different time requirements. All need to work in a concerted manner; however, the weakest element defines the total capacity. In order to be able to manage and plan airport capacities, the use of modelling and simulation has become the norm in airline planning and airline capacity management. Only by simulating the various airport processes can the airport's overall capacity be assessed. Simulation is also the way in which bottlenecks in the system can be identified.

Modelling means the translation of a process into a mathematical equation. This is useful for various reasons. First, with the right formulas we can extrapolate historical data to future predictions. Second, modelling can make us understand the relations between different elements. Third, with models that have proved to be valid, we can analyze what-if scenarios. It goes without saying that modelling and simulation have developed as indispensable tools for managing complex entities like airports.

Simulation is in bringing models to life. With the use of computers, we can simulate various models and analyze the interaction between them. Using this technique, we can identify bottlenecks in the entire system and find optimum capacities for all of the airport functions.

# PART III
# Aircraft

As this textbook is on aviation operations, we will not deal with the engineering dimension of aircraft. Many textbooks have been written on the technical aspects of aircraft. In Part III, we will look at aircraft as *operational capacity units* – the key role of aircraft for their operators – and as *capital goods*, perhaps the largest challenge for aircraft operations. In aviation operations, we do not look at aircraft as beautiful technical machines (which they are by the way); instead, we look at them as *capital intensive production units* that need to produce the connections that aviation operations offer to the world. They need to produce it *profitably*, both for its producer and for its exploiter.

Many do not consider the specifics of aircraft as relevant as all airlines operate the same aircraft, Boeing and Airbus. And this certainly is a very valid point to the extent that models from both manufacturers are comparable. Nevertheless, we need in aviation operations a clear understanding of the operational characteristics of aircraft. Aircraft differ profoundly from ground transportation vehicles as they need to conquer gravity.

We also need to understand that the structure of aviation is primarily defined by the technological characteristics of the aircraft. The way aviation has developed over time is foremost defined by technological developments. Aircraft technological developments over time have made aircraft more fuel efficient, resulting in increased payload capabilities and increased range capabilities. Aircraft have become more cost effective, enabling its operator to profitably operate aircraft against moderate fares. The tremendous growth of aviation in the last decades is therefore the result of technological developments. There is thus all the more reason to spend a part of this textbook on aircraft as production units and capital goods.

All these aircraft are developed, produced and supported by only a handful of suppliers, and many in aviation operations are either employed by a supplier or need to deal with suppliers, both on obtaining the equipment or on maintaining that equipment. We will therefore also look at the aircraft suppliers and on the structure of aircraft maintenance.

# CHAPTER 11

# Aircraft operation

Just as in the last part, let us start with the legal framework. To many students this is not the most appealing part of aircraft. But, an aircraft *cannot be understood* without understanding that it is certified, declared fit for safe operations, by authorities responsible for ensuring the safety of the complex machine that an aircraft is. We therefore start our journey into aircraft by looking at its certification.

## 11.1 Legal framework: aircraft certification

Certifying new commercial aircraft models is a tremendous responsibility for the certifying authority. The *NAA of the country of origin* of that new aircraft model is the authority that needs to fulfil this task. This NAA needs to issue the *Certificate* of *Airworthiness* (CoA). We discussed this in Chapter 2.

An airworthiness certificate is a *formal document that grants authorization to fly*. That formal document is the evidence that all rules and regulations have been met. Additionally, every individual aircraft needs an Airworthiness Review Certificate (ARC), which needs to be renewed every year. Without a valid CoA and an ARC, an aircraft cannot fly.

The NAA (National Aeronautical Authority) of the country where the *aircraft producer* resides is the authority that certifies the airworthiness of the aircraft. If done well and in full accordance with ICAO's annex 8, this certification will be followed by the NAAs of the countries where the aircraft will *operate*. Yes, all NAAs in the world need to certify their aircraft in order to fly with an airline in their country or land at an airport.

The FAA (Federal Aviation Administration) is the certifying entity for US-manufactured aircraft like Boeing, while EASA certifies aircraft from Airbus. The aviation agencies of other countries where aviation suppliers reside, like Canada, Brazil, Japan, China and Russia, follow the FAA and EASA in their certification processes. They do so for a strong reason: suppliers residing in these countries want to sell their equipment worldwide, so they need *market access* to the US and the EU. Therefore, their certification process needs to comply with, and often even needs the involvement of, either the FAA or EASA.

*National Aviation Authorities* are part of the national states, and have the powers to enforce rules on behalf of the state, by *withdrawing* a certification in case of a non-compliance with the rules, by imposing legal *fines* to non-compliant parties or even *prosecution* of individuals violating the rules.

### 11.1.1 Certification documents

A new aircraft type, its engines and all components installed in that aircraft need to obtain an *Original Type Certificate* (OTC) for being allowed to fly. When, in due course, the aircraft type is modified – for instance by installing another engine type – an ATC (*Amended Type Certificate*) is required. When modifications to the aircraft require only the installation of a specific system, and the aircraft type, as such, does not change, an STC (*Supplemental Type Certificate*) is required. All this is in conformity with ICAO annex 8.

All parts of the aircraft are certified under any of the above certifications. If, for whatever reason, an alternative supplier wants to produce an alternative part outside the original certification, a PMA (*Parts Manufacturing Approval*) is required.

When a design deficiency is discovered in an aircraft or one of its systems, an AD (*Airworthiness Directive*) note will be issued, ordering a modification or replacement of the deficient part within a specified time frame. If a design deficiency is considered a major risk for safety, the OTC can be withdrawn. The FAA did this in 1979 with the DC-10, in 2013 with the B787 and in 2019 with the Boeing 737MAX. All aircraft of the type of the OTC were *grounded, irrespective* of the nationality of the operator, or the location of the aircraft.

It is important to understand that all equipment, aircraft, engines and systems have *manuals* that describe how the equipment should be operated and how it should be maintained. These manuals are *part of the certification*, and need to be *approved* by the NAA. As such, operational procedures and maintenance processes, repair schemes in case of structural damage, are strictly regulated through the certification of manuals.

Obtaining an OTC requires *extensive data exchange* between the designing organization and the certifying authority, and requires extensive testing of the products under all thinkable conditions. The certifying authority needs to be *involved* in designing and testing by the design organization but always needs to remain at an *arm's length*. Certification processes are extremely expensive and form a significant part of total design cost of a product. The result of the efforts is a *generic certification* of a specific type of aircraft, type of engine or type of component. This is certification on *part number level.*

Not only does the hardware need to be certified, but also the organization that designs all that equipment needs formal approval to do so. They need a DOA (*Design Organization Approval*). Once equipment is in production, all companies involved in that production need to have a POA (*Production Organization Approval*).

## 11.2 The aircraft as production unit

An aircraft is a capital good that enables its operator to transport a certain amount of payload over a certain distance. This capability, the amount of *payload and the distance* that can be carried, defines the aircraft's *productivity potential.* The operator of the aircraft determines the *actual* productivity of the aircraft, based on the quality of airline planning, as we will see in Part IV.

An aircraft differs from other transport vehicles in the fact that an aircraft has to *conquer gravitation*. This means that it is all about *mass*. The mass of the aircraft itself and its payload are expressed in kgs or lbs. However, aircraft's range capability is also expressed in mass: the mass of the fuel necessary for covering the distance between the departure and the arrival airports with maximum payload. There is thus all the more reason to describe the various weights that we need to understand in order to gauge the aircraft's productivity.

### 11.2.1 Manufacturer's empty weight

First weight is the *Manufacturer's Empty Weight* (MEW). This is the weight of the base aircraft in certified flying condition, without interior, without livery and without eventual optional equipment that the operator has specified on the aircraft.

All individual aircraft of a certain type have the same MEW, or at least should have, making MEW a prime KPI for production consistency and quality of the aircraft manufacturer.

### 11.2.2 Operating empty weight

The second and very important weight is the *Operating Empty Weight* (OEW), which is often referred to as Dry Empty Weight by pilots. The OEW includes the MEW as well as the aircraft's interior, livery, optional equipment, safety equipment, catering inserts and the entire flight crew with their luggage. Thus, the OEW includes all of the aircraft, without payload and without fuel.

From the above follows that aircraft of the same type can have profoundly different OEWs. A low-cost operator has its aircraft with standard seats and minimum comfort amenities whereas a high-quality airline operates the same aircraft type with a luxurious interior installed, with flat-bed seats, huge TV screens and more. High cuisine catering is served on chinaware, adding to the high OEW typical for these airlines.

Aircraft Operating Empty Weights in tonnes

| | |
|---|---|
| A319 | 40,8 |
| A320 | 42,6 |
| A321 | 48,5 |
| A330-200 | 119,6 |
| A330-300 | 124,5 |
| A350-900 | 115,7 |
| A380-800 | 276,8 |
| | |
| B737-700 | 37,6 |
| B737-800 | 41,4 |
| B737-900ER | 44,6 |
| B747-400 | 178,8 |
| B777-200LR | 145,1 |
| B777-300ER | 167,8 |
| B787-8 | 110,1 |
| B747-8 | 191,1 |

**Figure 11.1** Aircraft empty operating weights

### 11.2.3 Maximum zero fuel weight

Next is the *Maximum Zero Fuel Weight* (MZFW), which is defined by the structural integrity of the aircraft. The MZFW defines the aircraft's maximum payload capability, as the difference between the MZFW and the OEW.

It is important to note here that aircraft built to transport passengers are almost never loaded up to the MZFW. The operator uses the volume of the fuselage for seating passengers, storing their luggage and filling up the remaining space with cargo, whereby the total weight of the actual payload almost never reaches the MZFW. Cargo operators on the other hand, and most notably the Parcel operators, often load the fuselage space with payload until the total weight reaches the MZFW.

The MZFW is aircraft-type specific: all aircraft of a certain type have the same MZFW; this weight is set by the authority that certified the aircraft. MZFW is a *certified* weight.

### 11.2.4 Maximum take-off weight

Also certified by the authority is the *Maximum Take-Off Weight* (MTOW). This weight speaks for itself; it is the *absolute maximum weight* at which the aircraft can take-off and climb out safely. It is identical to all aircraft of a specific type.

The difference between the MZFW and the MTOW is the weight that is available for fuel. This weight, together with the fuel efficiency of the aircraft, defines the aircraft's range capability.

It is very important to understand that the aircraft's fuel storage capacity, which is a *volume*, can only be used partly when the aircraft is loaded up to its MZFW. Therefore, an aircraft loaded up to its MZFW, loaded at maximum payload, is limited in range. The opposite is also valid; when the fuel tanks are entirely filled with fuel, the aircraft becomes limited in payload. This effectively means that an aircraft does not have a fixed payload and a fixed

Aircraft Max. Take-off Weights in tonnes

| | |
|---|---|
| A319 | 63,9 |
| A320 | 67,9 |
| A321 | 82,9 |
| A330-200 | 211,8 |
| A330-300 | 229,7 |
| A350-900 | 267,7 |
| A380-800 | 543,6 |
| | |
| B737-700 | 70,0 |
| B737-800 | 78,9 |
| B737-900ER | 85,0 |
| B747-400 | 412,2 |
| B777-200LR | 347,0 |
| B777-300ER | 351,1 |
| B787-8 | 227,6 |
| B747-8 | 447,1 |

**Figure 11.2** Aircraft max take-off weights

range capability. Every individual operator chooses a *balance* between payload and range, often even per destination. When an airline has a strong position on a certain route in both passengers and cargo business, the airline may use a B777-300ER for that route. The same airline operating passengers only can cover the same distance with an A330-300.

## 11.3 Payload-range diagram

When we want to know the productivity potential of an aircraft, its payload-range capability, we need to read the payload-range diagram of that aircraft. Therefore, for understanding an aircraft's *productivity potential*, we need to understand this graph. Also, for comparing various aircraft types, the payload-range diagram is indispensable for aviation operations professionals. Let us examine how the payload-range diagram is constructed.

The figure below shows the first step in the build-up of the payload-range diagram. Vertically on the left are the aircraft's weights. The X-axis of the diagram starts at the OEW of the aircraft. If we were to load this aircraft up to its MZFW, we would get the max payload line, representing the aircraft's actual range capability at maximum payload. Here, line A represents the aircraft's fuel burn, with angle α representing the aircraft's fuel efficiency. The point where the fuel burn line reaches the MTOW of the aircraft represents the aircraft's max range when loaded up to the MZFW.

It is important to understand that the trip fuel as represented in line A includes the necessary *reserve fuel*: fuel for *holding* at the destination airport; fuel for flying to an *alternate* airport; fuel for an *overshoot*; and, depending on the distance flown, *contingency* fuel as a percentage of the net trip fuel. Furthermore, the fuel required as presented by line A assumes *zero winds* en route.

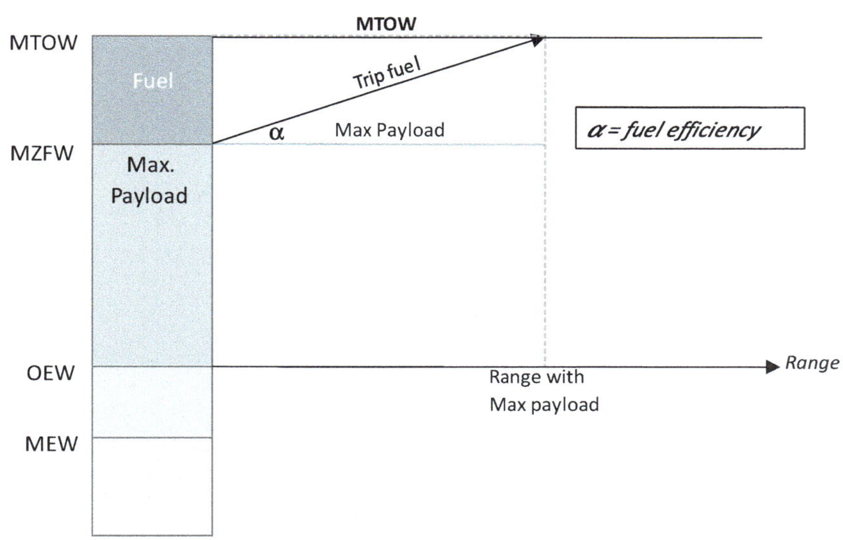


**Figure 11.3** Max payload line

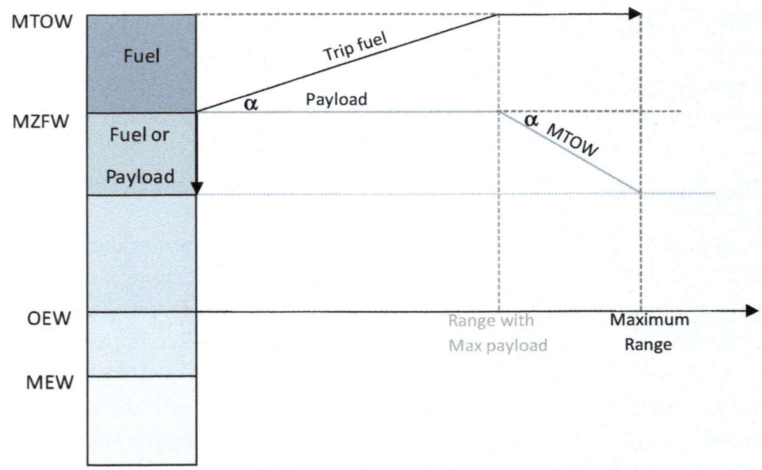


**Figure 11.4** Payload versus range trade-off

When we want to fly further with this aircraft, we need to lower its payload, allowing more weight for fuel. Hence, the payload line needs to go down.

The trip fuel line goes down, whereby the angle α by which is goes down is *identical* to the angle of the fuel burn line. Within this part of the trip fuel line, the aircraft will depart at its MTOW; this is why we call this the MTOW line. Here we see graphically the actual trade-off between payload and range.

If required, we can even fly further with the aircraft. As the aircraft's actual range capability is among other things also a function of the aircraft's mass, we can lower this mass by further lowering its payload.

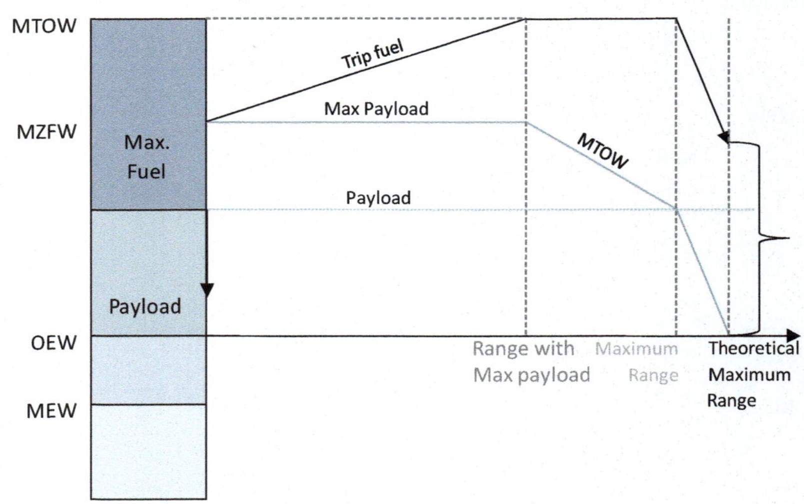


**Figure 11.5** Maximum range

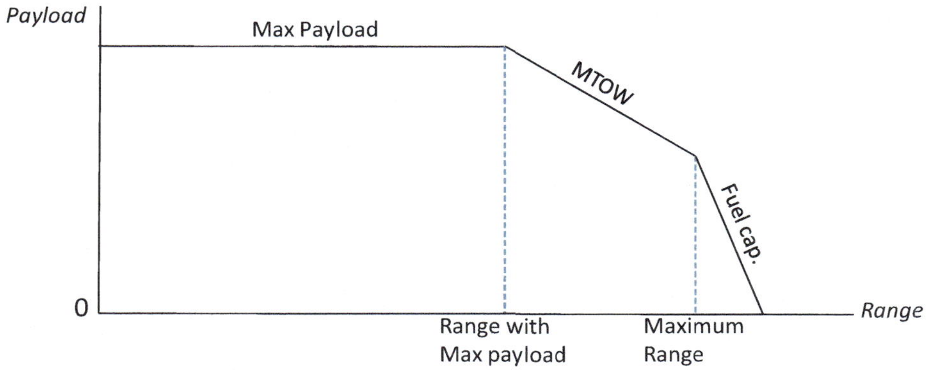


**Figure 11.6** The payload-range diagram

By lowering the aircraft's payload, whereby the aircraft will take-off with an actual take-off weight lower than its MTOW, we can further increase the aircraft's range, up to the theoretical max range, at which the payload is zero.

If we understand the above, we can now delete all help lines and read a payload range diagram.

If we now plot the actual payload of a specific airline, we can read the range capability of the aircraft with that specific payload.

If we plot various aircraft types in the same diagram, we can now compare these various types, like the various variants of the Boeing 777.

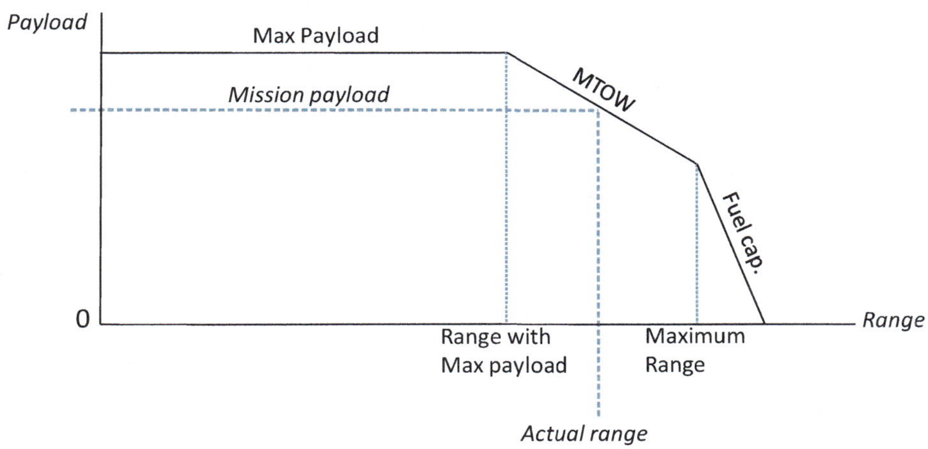


**Figure 11.7** Actual range at mission payload

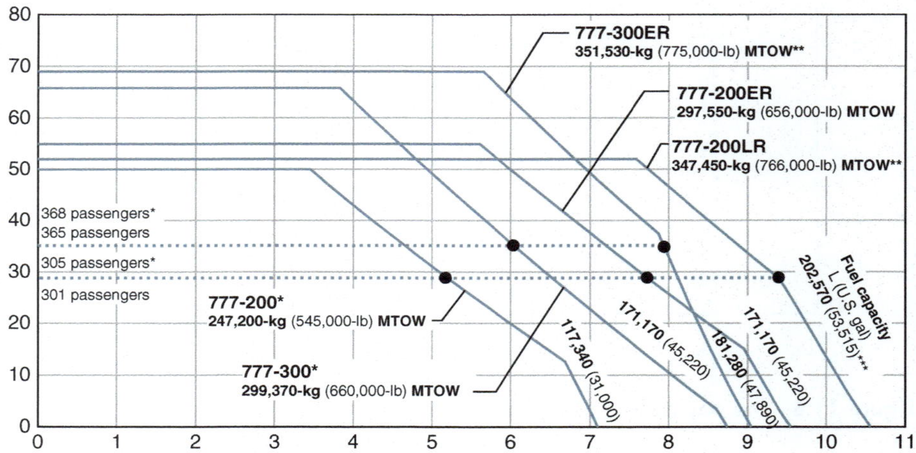


**Figure 11.8** Payload-range diagram B777

## 11.4 Aircraft field performance

The payload-range capability of an aircraft, as expressed in its payload-range diagram, contains a very important assumption: it assumes that the aircraft can take off at its MTOW, an *unrestricted operation*. This is the case if the runway is long, wide and strong enough; the airport elevation is not too high; the outside air temperature is acceptable; and there are no high mountains in the vicinity of the departure airport. If one of these conditions is not met, the aircraft might face *a limitation in its allowable TOW*, limiting the actual payload-range capabilities, limiting the aircraft's productivity.

### 11.4.1 Runway dimensions

Let us start with runway length. We discussed airport runway capacity in Chapter 10. An aircraft needs, by ICAO Annex 6 ruling, to be able either *to reject its take-off run, or to continue its take-off with one engine failing at take-off.*

Large, wide-body aircraft need runways with a Take-Off Run Available (TORA) length of some 3,000 m and a 45 to 60 m width for unrestricted take-off. Furthermore, such a runway needs to be strong enough to carry the heavy loads of the aircraft departing at MTOW. Narrow-body aircraft can operate unrestricted on runways with a length of approx. 2,200 m.

From the above it follows that big aircraft can only operate from and to large airports, while narrow-body aircraft can operate between smaller regional airports as well.

### 11.4.2 Outside air temperature

The actual thrust performance of an aircraft jet engine goes down with increasing outside air temperatures. To understand why this is the case we need some basic understanding of how a jet engine works. The engine sucks air and compresses the air to a very high pressure,

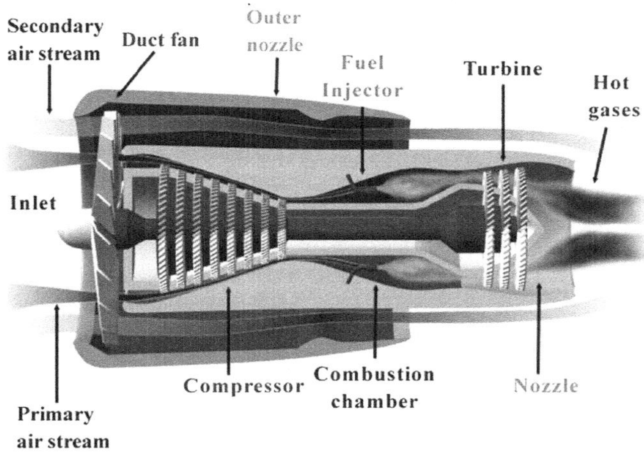


**Figure 11.9** The high-bypass engine

whereby the *temperature of the air increases.* This compressed air is than combusted by adding fuel, in the combustion chamber of the engine whereby the air expands. *This expansion generates the engine's thrust.* The amount of expansion depends on the difference between the air temperature before and after the combustion. Therefore, *the hotter the intake air is, the lower is the resulting thrust.*

Now, up to a certain point this decline in thrust can be compensated for by increasing the power setting of the engine. But at a certain outside air temperature the compressed intake air becomes too hot for the materials of which the engine compressor is built. From this point, we cannot further increase the thrust setting of the engine. This point is called the *engine's flat rating.* Engine manufacturers often offer H+H (hot-and-high) variants of their engines with higher flat ratings; these engines are more expensive to operate and maintain. But an airline operating from and to airports with high outside temperatures will benefit from this extra cost.

In order to standardize this subject, ICAO has defined ISA (*International Standard atmosphere*), whereby 15°C at sea level is called ISA, going up one point for every degree Celsius. So, an outside air temperature at sea level of 30°C is called ISA+15. And this is the flat rating for most modern aircraft engines.

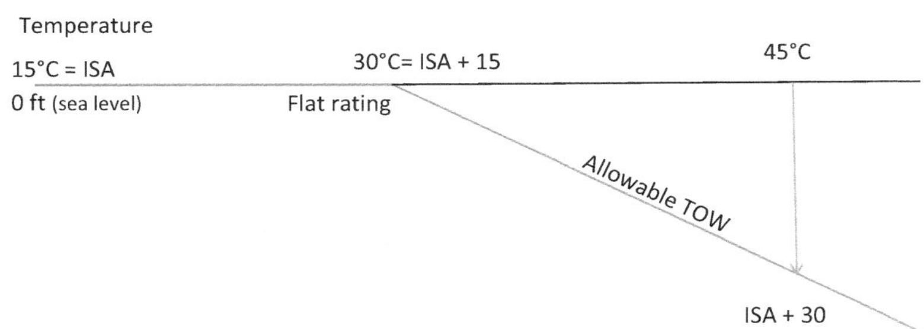


**Figure 11.10** ISA and ambient temperatures

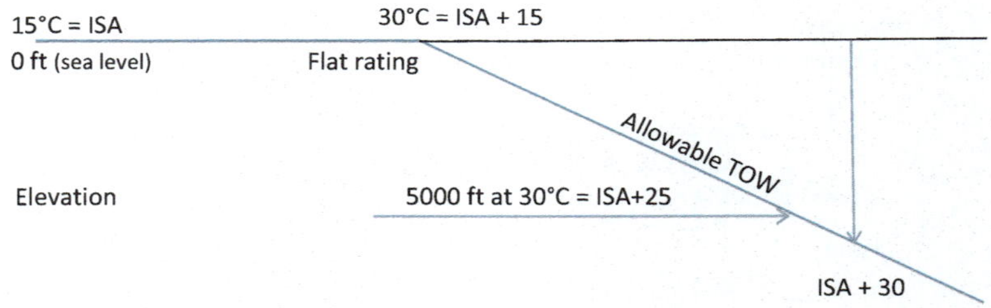


**Figure 11.11** ISA and airport elevation

## 11.4.3 Airport elevation

At higher altitudes the air gets thinner; we encounter this phenomenon when we are high in the mountains. It means that at higher altitude the air pressure drops as the air contains less oxygen atoms. The air that is combusted in the engine is actually the oxygen. Therefore, less oxygen in the atmosphere means less thrust produced by the engine. As the effect is the same as with high temperatures, the elevation of a departure airport can also be expressed in the ISA system, whereby every 500-ft elevation translates into one ISA point. So, an airport with an elevation of 5,000 ft at 15°C is ISA +10.

When high temperatures at high altitudes come together, the deterioration of engine power, and therefore, the deteriorating allowable TOW of the aircraft, is compounded. The level by which an aircraft is capable of dealing with this situation is called the aircraft's hot-and-high performance. Professionals in the US call it the aircraft's "Denver performance" as Denver CO is situated at a high elevation with high temperatures in the summer season.

## 11.4.4 Runway bearing strength

Have you ever wondered why large, long-range wide bodies have so many wheels? The answer is quite simple; this is to reduce the wheel pressure of the aircraft on the apron-taxiway and runway of the airport. Earlier in this chapter we presented MTOW of contemporary aircraft, and indeed, an aircraft at MTOW contains an impressive mass. When looking at the largest aircraft in the industry, the Airbus A380, we see that this aircraft can take-off with a mass of some 550 tons. When we know this, we understand why this aircraft is equipped with a landing gear containing 20 wheels!

Here also ICAO has introduced a standard: the ACN (the *Aircraft Classification Number*), whereby the load per wheel (MTOW in tons divided by the number of main wheels) x2 results in the aircraft's ACN.

Airports need to establish the actual strength of their paved airside surfaces, expressed in PCN (Pavement Classification Number); for a regular operation an aircraft's ACN needs to be lower than the airport's PCN. Most notably in Asia many airports are constructed on weak soil, resulting in a low airport PCN, at some locations severely limiting the productivity potential of aircraft operating at such airports.

So, in conclusion, the maximum allowable take-off weight of an aircraft is dependent upon:

**Figure 11.12** A380–800 main gear

- The structural limitation, which is set by the manufacturer;
- The available runway length (the shorter the runway, the lower the allowable take-off weight);
- The presence of obstacles requiring a larger than average climb performance after take-off (the steeper the climb that is required, the lower the allowable t/o weight);
- The outside air temperature (the higher the temperature the lower the allowable t/o weight);
- The ambient pressure (the lower the pressure, so the higher the airport's elevation, the lower the allowable take-off weight); and
- The bearing strength of the airport pavements.

## 11.5 Turnaround characteristics

So far, we have discussed the aircraft's productivity potential as a flying machine. However, aircraft need to spend time on the ground as well. Passengers and cargo need being loaded and unloaded, fuel need to be loaded on the ground, at the airport. There is thus all the more reason to look at the turnaround characteristics of the aircraft. How much time and effort is needed to turn the aircraft around at the airport. It is important here to remember what was discussed in Chapter 9 on airport services. Here we saw that ICAO categorizes airports on the basis of maximum aircraft sizes and capacities. An aircraft needs to be *compatible* to an airport category in order to operate to that airport. And indeed, big aircraft can only operate on big airports.

The speed by which passengers can deboard the aircraft upon arrival is defined by the *number of available passenger doors* of the aircraft, and the *number of aisles* in the cabin. Boarding and deboarding of passengers occurs at the aircraft's left-hand side, while servicing the cabin is done at the right-hand side of the aircraft. Twin-aisle wide-body can deboard faster on a per passenger base than single aisle narrow-body aircraft. The same is true of the boarding process upon departure of the aircraft. Also relevant, and in close conjunction, is *the floor height of the cabin*. We need either stairs or avio-bridges to deboard and board passengers.

The aircraft's height is also important for unloading and loading luggage and cargo. We may need conveyor belts or lift devices to cover the height. Aircraft become higher as the modern fuel efficient engines are equipped with a fan with a large diameter. As the fan becomes larger, the aircraft need to stand higher on its landing gear to ensure clearance between the engine cowling and the surface.

Another important element of an aircraft's turnaround characteristic is the use of *containers* for luggage and cargo. Wide-body aircraft use containers, lowering the amount of time needed for unloading and loading the aircraft. Most narrow-body aircraft do not use containers. The exception is the Airbus A320 on which aircraft belly containers are an option. However, these containers are aircraft type specific. In order to allow a cost-efficient container logistics, these containers need to be *standardized*.

The aircraft's galleys are loaded thru designated service doors or passenger doors at the right-hand side of the aircraft, and close to the galley inside the aircraft. Often, these service doors are located just after the wing, whereby the manoeuvring space for the catering truck is limited, with potential risk of colliding with the aircraft's flap tracks.

Also, the lower deck cargo hatches are located at the right-hand side of the aircraft. Their position may interfere with the cabin doors, in that access to both doors simultaneously is not always possible. Fuelling as well as water and waste servicing is usually done from the aircraft's left-hand side and may interfere with passenger handling. If fuelling is done from the right-hand side (which is also possible) it will interfere with cargo- and catering operations.

More in general, an aircraft's turnaround characteristic is largely defined by the extent to which the various turnaround tasks can be performed simultaneously. This is all dependent on the layout of the aircraft.

## 11.6 Noise characteristics

Aircraft are fairly noisy machines, certainly in the take-off and climb out phase of a flight, even if aircraft noise has reduced spectacularly over the years. This noise is related to the engines, and as different aircraft types fly with the same engine type, noise characteristics of aircraft are comparable within a certain generation of aircraft. Noise characteristics therefore can be a reason for operators to replace their fleet with a newer aircraft type, most notably a *new engine generation*.

Aircraft noise is generally expressed in dB(A): decibels corrected for the sensitivity of the human ear. The decibel is in itself a complicated unit as it is not linear but exponential, whereby an increase of 10 dB(A) is an increase in noise by a factor of 100.

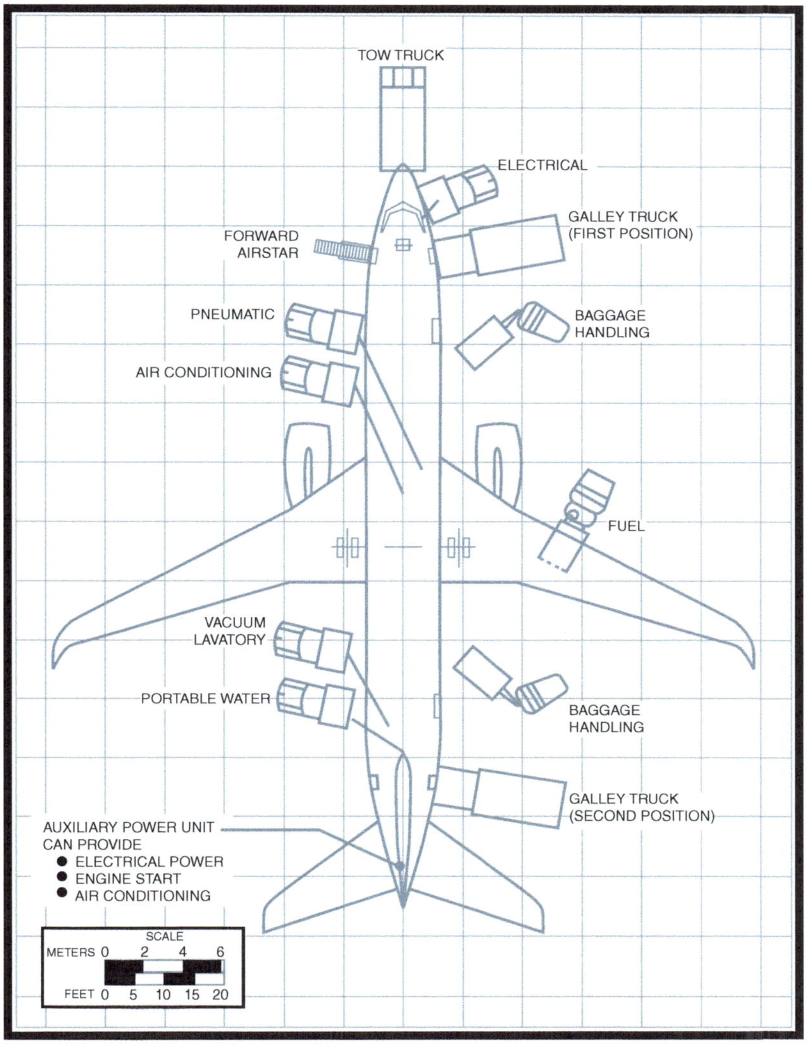


**Figure 11.13** Parallel turnaround processes

Aircraft can be compared on noise characteristics by comparing their 85dB(A) noise footprint, the area in which an aircraft produces 85dB(A) or more during take-off and climb out. It is not only the nominal noise of an aircraft but its *climb characteristics* that define its 85dB(A) noise footprint. This complicates matters further. When a noise footprint of an aircraft is presented, first question is what the assumed TOW of that aircraft is, as its actual weight defines its climb ratio, and therefore the size of its noise footprint. Especially freighter-aircraft, taking off at MTOW, can have a larger noise footprint than generally regarded for the aircraft type.

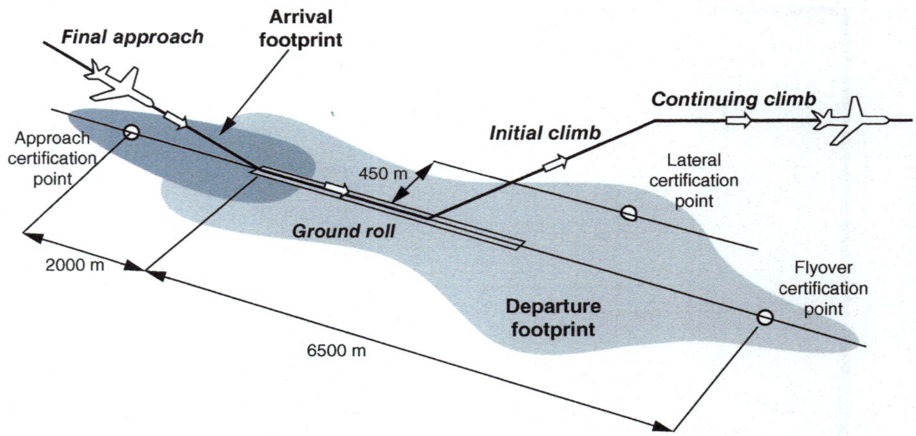


**Figure 11.14** Noise footprint

## 11.7 ETOPS

Most contemporary aircraft types are equipped with two engines. This implies that in case of an in-flight engine failure, the aircraft needs to be able to fly on one engine only. Indeed, these aircraft can fly on one engine. For how long it is allowed to fly on one engine only, is defined by its ETOPS certification. It stands for Extended-range Twin Operational Performance Standards. ETOPS-90 means that the aircraft, its engines and its operator are jointly certified to fly with that aircraft on one engine for a maximum of 90 minutes. Today's twin-engined aircraft types come with ETOPS180 or even more. But whatever ETOPS standard the aircraft possesses, the operators need to comply with a whole set of rules and regulations to make use of that capability. Maintenance procedures should be ETOPS-compliant and differ from standard maintenance; pilots should have undergone special additional training and be kept current on ETOPS requirements and the flight ops; or dispatch organization should be ETOPS compliant with additional training, recurrent training and qualification of its personnel. In summary, flying extended range with twin-engine aircraft is very economical and makes great sense, but it comes at a price. To put the whole issue in perspective ETOPS was originally based on the assumption that engine reliability was the crucial factor. However, it isn't anymore. Engines are much more reliable than they used to be and ETOPS is becoming more and more an obsolete concept. Rather, the whole idea of flying over remote areas where there are no or very limited diversion airports available has become a subject of scrutiny, independent of number of engines.

## Suggestions for further reading

Mohammad H. Sadraey: *Aircraft performance*. Routledge, 2006.

CHAPTER 12

# Aircraft economics

If we consider aircraft, we primarily concentrate on its technical characteristics. However, in order to understand aviation operations, and specifically the financial and economic aspects of aviation operations, we need to look at aircraft from a different angle: the aircraft as a capital good and a machine for cost-effective production. We will therefore take a closer look at these aspects in this chapter.

## 12.1 The aircraft as capital good

Aircraft are expensive machines to acquire. List prices start at approximately $30 million for a 70-seat Embraer 175 and go up to over $400 million for the largest unit on the market: the Airbus 380-800. But as aircraft are *capital* goods, it is not how much an aircraft costs but *how much profit it can earn* for its owner which determines whether it is a sound investment. And indeed, the larger the aircraft, the more expensive it is to acquire and the higher the *earning capability* becomes. And that, the earning capability, defines the *value proposition* of an aircraft, not its purchase price.

Aircraft models in the market today have tremendous earning capabilities. If operated well in the fleets of efficient airlines, aircraft can earn the investment back in about ten years' time. Given the average life span of the aircraft of say 25 years, aircraft can over time indeed generate healthy *Returns on Investment* (ROI), making aircraft attractive objects for investment entities like banks or institutional investors.

The combination of the two aspects described above, the high prices and the sound ROIs, have resulted in the situation that the ownership of the aircraft and the operation of the aircraft are often *separated.* In aviation, we see *owners* of aircraft and *operators* of aircraft. Aircraft owners often are finance providers like banks or leasing companies. Aircraft operators are airlines. We will deal with them in Part IV of this book, where we will see that actual ownership cost per flight hour does not so much depend on the price of the aircraft but on the number of annual flight hours that the aircraft is scheduled to fly and the *planning efficiency* of the airline operator.

List prices per 1-1-2017 for various aircraft

| | | | |
|---|---|---|---|
| B737MAX7 | 92,2 | A319NEO | 98,5 |
| B737MAX8 | 112,4 | A320NEO | 107,3 |
| B737MAX9 | 119,2 | A321NEO | 125,7 |
| B7777-300ER | 347,1 | A330-900 | 287,7 |
| B787-8 | 229,5 | A350-800 | 272,4 |
| B787-9 | 270,4 | A350-900 | 308,1 |
| B787-10 | 229,5 | A350-1000 | 355,7 |
| B747-8 | 386,8 | A380-800 | 432,6 |

*Here we see published list prices for aircraft in US$ per 1-1-2017*
*Actual prices upon delivery depend on options selected, actual delivery date, as the prices will be escalated, and last-but-no-least, the bargaining power of the buyer.*

**Figure 12.1** List prices for various aircraft

### 12.1.1 Aircraft life cycle

In the paragraph above we stated that the average life span of an aircraft is roughly 25 years. However, individual aircraft types can divert significantly from this average. A successful aircraft type can remain in operation for up to 35 years, like the DC-9 in the fleet of Delta. An aircraft type that does not meet expectations, like the McDonnell Douglas MD-11, was disposed of by many airline fleets within ten years of operation. So, what defines the success of an individual aircraft type?

Success is a combination of various factors: the aircraft's payload-range capability, its operating cost, its ease of maintenance, is overall reliability and last but not least, the quality of the support provided by its producers.

When expressed in time, say years, the technical life span of a long-haul aircraft is longer than of a short-haul aircraft. Reason for this is the fact that the *technical life span* of an aircraft frame is 60,000 cycles, whereby a *cycle is a single flight irrespective of the flight duration*. Taking off, climbing to high altitudes with low pressure, descending and landing back on the higher-pressure surface generates *metal fatigue* on the aluminium airframe. This is the limiting factor for an aircraft's technical life. As a long-haul aircraft makes just a few cycles per day, and a short-haul aircraft makes many cycles in that same day, the long-haul aircraft lasts longer over time. The 60,000-cycle boundary defines the *technical obsolesce* of the aircraft.

The boundaries for *economic obsolesce* are harder to define. Economic obsolesce occurs when a new aircraft type provides clear operational, technical and/or economic advantages over the older type. This is partly defined by the technical characteristics of the aircraft, but also by the market strategy or fleet strategy of the operator. Some airlines take a new aircraft type in their fleet as soon as possible in order to gain from the new technologies applied, other airlines remain happy with existing aircraft types for many years gaining on the experience they have with the type and enjoying the low ownership cost that come after many years of operation. As new aircraft types provide better fuel efficiency, the actual fuel price is a major factor in determining the economic obsolesce of an aircraft. At low fuel prices, the low ownership cost that comes with older aircraft can outweigh the higher fuel consumption, certainly at operators with a professional maintenance outfits with tons of experience and spare parts for the old aircraft type. But one can also argue that they dance on the side of the volcano, to be hit hard when fuel prices rise. Economic obsolesce is indeed hard to define.

### 12.1.2 Aircraft values

Actual values of aircraft in service clearly express the market success of an aircraft type. Successful types for which demand is high retain their market value for many years. Here the value of the used aircraft is in its imminent availability against the manufacturers' lead-time for a new aircraft of that type. On the other hand, disappointing aircraft types can lose their market value very rapidly, even to the extent that when no second-hand buyer is interested, the aircraft value can go down to its scrap value. When an airline disposes of these aircraft, they are stored in the desert, hoping for better times but likely waiting for the scrap hammer.

A special category here is formed by the very large aircraft. These aircraft, like the Airbus A380, are in operation at very large airlines with strong market positions. Such airlines prefer phasing in new aircraft of a type with minimal interest in second-hand propositions. This implies that second-hand market values for very large aircraft are low, not because of obsolesce or disappointing performance, but solely because of a lack of second-hand buyers.

In addition to the generic market value of an aircraft type, individual aircraft of the same type can differ in value as well. This is mainly defined by the *quality of maintenance and repair* on the individual aircraft. A poorly maintained aircraft, or an aircraft repaired after a serious incident, will have a lower value than an untouched aircraft well looked after.

All of the above indicates that investing in aircraft can contain *more risk exposure* than some investors are aware of. Investors can generate a very healthy return on their investment in aircraft, but can also lose their money. The German pension fund financed the first A380s for launch, and customer Singapore Airlines enjoyed a healthy return during the ten years that the airline leased the aircraft. Now, after the termination of the lease the same pension fund find itself stuck with aircraft for which there is no second-hand demand.

## 12.2 Aircraft financing

Aircraft are high-value capital goods, and acquiring aircraft by airlines results in the need for finance, provided primarily by banks. Banks finance the sales transactions between the suppliers and the airlines. The large multinational bank consortia are the prime finance providers of aviation. These banks have dedicated aviation branches within their organizations.

Finance by banks come basically in two ways. First, there is the straightforward *loan*, whereby the capital good is the collateral for the loan, a mortgage. The loan itself comes against a market conform interest, but as aviation investments are generally regarded to be moderately risky by banks, the interest rates are moderate as well. In case of default by the loaner, the aircraft becomes at the disposal of the bank, and this is why banks assume their risks to be moderate. Successful aircraft types retain their value for a long time, and replacing the aircraft at a new operator after a defaulted first operator is assumed to be relatively easy as aircraft can be placed anywhere in the world, a quality that real estate, for example, does not possess.

The second, very common, form is the *financial lease*. Here, the bank is the owner of the investment, and the lessee, the airline, pays a monthly rental fee. At the end of the lease period *the lessee becomes the owner* of the capital good. The financial lease is very popular amongst airlines as the construction provides the finance departments of airlines flexibility on balance sheet policies. For airlines with a weak balance sheet – a low solvability – financial leasing is often the only viable option for fleet renewal. This does however not imply that leasing has no effect on the balance sheet of the lessee, as contemporary accounting regulations stipulate that the entire lease obligation needs to be expressed on the balance sheet.

Leasing companies can also provide financial leases, but their main business is in *operational lease* of aircraft. Many airlines cannot finance or, more common, do not want to own the aircraft that they operate. These airlines want *flexibility* in fleet strategy, and this is exactly what the leasing companies offer as finance providers. In an operational lease, the airline leases the aircraft against a monthly lease rate. At the end of the five- to ten-year lease period the aircraft *returns to the leasing company* as the owner of the aircraft.

Contrary to the banks, leasing companies place large orders for aircraft in single configuration at the aircraft suppliers. These transactions come with significant discounts on the catalogue price of the aircraft. The leasing companies then actively market these aircraft in order to find an airline willing to lease it when it comes off the production line. Lease rates are based on the catalogue price of the aircraft and it is in the discounts in the aircraft purchase price where leasing companies earn part of their profits.

The other source for profit at lease companies is in the difference between the actual market value of the aircraft at the end of the lease period, against the book value of that aircraft on the balance sheet of the lease company. This makes lease companies as finance providers distinctly different from banks. Contrary to banks, lease companies closely monitor their aircraft in operation at lessor airlines. They stipulate in great detail in their contracts with airlines how the aircraft should be operated and how it should be maintained and repaired if necessary. In order to ensure this proper maintenance, the lessee needs to pay a monthly *maintenance reserve*, from which the maintenance is paid. Above all, they stipulate the return conditions for the aircraft at the end of the lease period. This is all meant to ensure a high market value for the aircraft after the termination of the lease contract, which is often valid for a ten-year period.

Where the aviation branches of banks mainly employ finance people and perhaps a research desk, leasing companies require engineering knowledge as well. They employ engineers or outsource their need to specialized engineering organizations that monitor the airline's operation of the aircraft on behalf of the leasing companies. Yes, leasing companies provide work for many engineers.

## 12.3 Aircraft types

A few decades ago aircraft came in various shapes and with various concepts, but since then the commercial aircraft has evolved into mainly one form: the aircraft with two engines hanging under the wing. Two engines only has become feasible for wide-body aircraft as well due to the use of (Ultra) High-Bypass Fanjet engines. Placing them under the wing, or actually slightly forward from the wing, in the centre of the aircraft, allows the fuselage to be extended or shortened without changing the aerodynamic chord or the centre of gravity for the aircraft – the technical prerequisite for operating one aircraft type in various fuselage length, in various seating capacities.

Improvements in engine fuel efficiency comes mainly from deploying a larger fan, we therefore see new engines appearing at higher diameters of the engine cowling. Aircraft with big fans stand higher on their gear on the ground to ensure ground clearance for these bigger engines. Besides the large fan, engines gain fuel efficiency and lower emissions by higher operating temperatures. The lowest emissions caused by burning fuel occur when the burn process is stoichiometric; at that point exactly all fuel molecules 'burn' with the available oxygen. No fuel molecule is left unused. Unfortunately, the temperature that is reached at the stoichiometric burn is so high that as a side effect the nitrogen in the air reacts with oxygen, thereby forming NOx pollution and emission. A tremendous engineering effort is put in reaching the goal of complete burn of fuel (no fuel molecule left unburned), but at a temperature which doesn't lead to the forming of NOx.

From Sections 11.2 and 11.3, we learned that an aircraft's range is not defined by its fuel tank capacity, but by *the weight of the fuel.* Now we can understand why aircraft manufacturers have always tried to lower the fuel consumption of their aircraft. Fuel efficient aircraft are not developed because fuel is expensive, but because it *increases the productivity potential of the aircraft.* Less fuel per mile means *more range*, less weight for the fuel means *more payload.* So, the main challenge of a gallon of fuel is not its varying price but its constant weight of 6.7 lbs.

Fuel efficiency is actually the *first and paramount parameter* by which an aircraft is designed. If we want an aircraft with a long range, we need a wing that can store the required fuel. So, given a certain quest for range, fuel efficiency of the engines defines the necessary *size of the wing*, and this defines the optimum size of the fuselage. This is why long-haul aircraft are big, and short-range aircraft are small. This is why the industry does not provide small commercial aircraft with a large range capability, nor mega aircraft with short range.

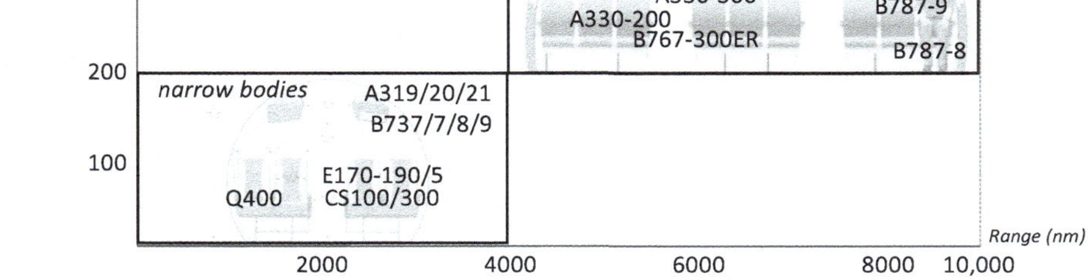


**Figure 12.2** Passenger capacity and range for various aircraft

These big, long-haul aircraft have *good economics on a per seat* basis. But they also have many seats which need to be filled. The airline industry needs long-haul aircraft that are economic with less seats to fill, and this quest can only be fulfilled by *lowering fuel consumption*. This lowers the space required for the fuel resulting in a smaller wing and consequently a smaller fuselage.

Less weight for the fuel also means that the structure of the airframe can be built lighter as it needs less weight to carry. The industry now has come to a point by which the more compact airframe can be built with lighter materials like composites, lowering the MEW of the aircraft. As fuel efficiency is also defined by aircraft mass, the lower MEW translates into improved fuel efficiency as well.

## 12.3.1 Narrow-body aircraft

When we look at today's narrow-body aircraft, we see that the application of more fuel-efficient engines on an existing airframe results in relatively small aircraft capable of either covering longer distances, or carrying more payload than its predecessor. Indeed, the latest version of the Boeing 737, the MAX, as well as the latest Airbus, A320 NEO, can carry a full passenger load over the North Atlantic or can carry well over 200 passengers on continental routes. The smallest capacities also, like the Emb175 or the A220-100, have range capabilities up to 3,500 nm.

The above can be regarded as a shift, as narrow-body aircraft are operated on short- and medium-haul routes. Seating capacities of narrow bodies range from 75 on the A220-100, up to 237 on the A321NEO. Indeed, capacity ranges, as narrow-body aircraft types come as family types, whereby the aircraft type is available with various fuselage lengths, resulting in various capacities and corresponding OEWs. These capacity ranges are optimal for short- and medium-haul operations. When an airline wants to increase capacity, it must primarily do that by increasing frequencies, as short-medium haul markets are predominantly frequency driven.

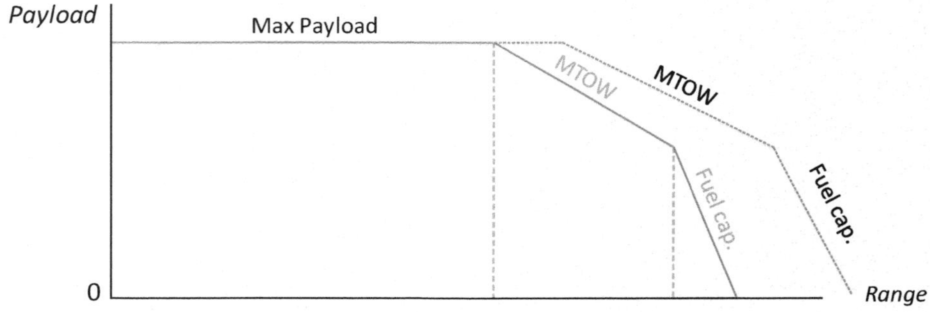


**Figure 12.3** Improved fuel efficiency in B737MAX and A320NEO

### 12.3.2 Wide-body aircraft

These large capacity aircraft are deployed on the long-haul routes, albeit that many airlines use these large capacity aircraft on shorter routes also when they are confronted with slot restrictions at the destination airport.

Whereas traditionally long-haul aircraft were powered by four engines, the twin configuration is applied to most contemporary long-haul aircraft types. Wide bodies come in various sizes as the narrow-body models, but there is a difference in the flexibility by which these long-haul family types can be deployed. A larger fuselage means more weight. On short- or medium-haul operations, weight restrictions do not occur, but on long-haul operations the weight differences between the various family aircraft result in different range capabilities as well, whereby in general the shortest variant comes with the highest-range capability. This can be compensated for by installing more fuel capacity in the larger version of the type.

### 12.3.3 Freighter aircraft

Dedicated freighter aircraft are almost always variants of passenger aircraft. Such aircraft can be newly built as freighter aircraft, or be converted from an original passenger aircraft. The former category provides the advantage of an aircraft optimized for cargo operations. The latter category accepts the concession of having a higher OEW than necessary, but provides the advantage of lower ownership cost as the aircraft is already depreciated by the previous passenger operator. Freighter conversion can generate a lively second-hand market for passenger aircraft that are regarded economically obsolete for passenger operations.

Not every passenger aircraft type is suitable for efficient cargo operations. An effective freighter aircraft has an optimum balance between weight regime, most notably its MZFW, and fuselage volume, whereby the optimum of this balance is defined by the weight density of the cargo. Cargo is, on a square metre basis, much heavier than passengers. If a cargo operator loads its aircraft up to the MZFW, the fuselage volume should be fully used. A fuselage optimized for passenger operations may well be too large in volume for cargo operations. This is well illustrated by Boeing's dedicated freighter, the B777-200F. This aircraft has the weight regime, the engines and landing gear of the B777-300ER, but all this in combination with the fuselage of the shorter B777-200.

In cargo operations, dedicated freighter aircraft offer advantages over belly cargo operations on passenger flights. The dedicated freighter can fly to the optimal cargo airport and transport large cargo and dangerous goods, whereby the operations are controlled by a dedicated cargo airline and its freight forwarders.

Loading the aircraft up to its MZFW limits the distance that the aircraft can fly as we saw earlier. For this reason, cargo aircraft make intermediate fuel stops en route, where the passenger aircraft flies non-stop. Setback of this kind of operation is that the fuel stop generates additional cycle cost and accumulates cycles to the airframe. Therefore, a passenger aircraft should not be that old for a cost-effective conversion to a freighter aircraft.

## 12.4 Economic characteristics

Every aircraft type has its economic characteristics. These are the result of various factors. First factor is the fuel efficiency of the aircraft type, resulting from the specific fuel consumption of the engines and the aerodynamic efficiency of the wing. Closely related to that is the second factor, the weight regime of the aircraft, especially the ratio between the aircraft's OEW and MTOW. We will discuss this in the next paragraph.

The third factor defining the economic characteristics of an aircraft is its technical complexity. This factor defines the maintenance requirements, the maintenance cost, of the aircraft. Related to this is the fourth factor, the number of aircraft in service. Popular aircraft types, in service at operators all around the world, have the advantage of many maintenance providers being familiar with this aircraft type, and with spare parts widely available.

Regarding the economic characteristics of aircraft types, we need to realize that it is not so much the aircraft type itself, but merely the way that the aircraft is operated by an airline that defines its economics. We will discuss this in great detail in Part IV.

### 12.4.1 Cycle cost

As said, the weight regime of an aircraft defines its economic characteristics. We saw that long-range aircraft are heavy and short-range aircraft are light. Long-range aircraft need to be large and strong in order to carry the fuel required for a long-range flight. The quest for strength implicates that a long-range aircraft is, on a per seat basis, heavier than a short-range aircraft.

In order to fly economically, an airline needs to use the specific qualities of an aircraft type. Thus, a long-range aircraft needs to fly long-range routes in order to be economical. If an airline would use a long-range aircraft for short routes, that airline pays for the long-range capability without using it.

| | OEW | seats | OEW per seat (kgs) |
|---|---|---|---|
| A319 | 40,8 | 129 | 316 |
| A320 | 42,6 | 157 | 271 |
| A321 | 48,5 | 192 | 253 |
| A330-200 | 119,6 | 253 | 473 |
| A330-300 | 124,5 | 295 | 422 |
| A350-900 | 115,7 | 314 | 368 |
| A380-800 | 276,8 | 525 | 527 |
| | | | |
| B737-700 | 37,6 | 138 | 272 |
| B737-800 | 41,4 | 175 | 236 |
| B737-900ER | 44,6 | 190 | 235 |
| B747-400 | 178,8 | 416 | 430 |
| B777-200LR | 145,1 | 320 | 453 |
| B777-300ER | 167,8 | 370 | 454 |
| B787-8 | 110,1 | 230 | 473 |
| B747-8 | 191,1 | 467 | 409 |

**Figure 12.4** Aircraft operating empty weights per seat

Aircraft generate cycle cost: the total cost to take-off an aircraft, climb it to cruise level, and to land it at the destination airport. These cycle costs are high for heavy aircraft and low for small aircraft. Heavy aircraft need a lot of engine power and thus more fuel for take-off and climb. Heavy aircraft have many brakes required for landing. Its large engines need expensive maintenance on a per cycle basis and generate high landing charges on the arrival airport. Therefore, as a rule of thumb it is fair to conclude that due to its weight specifics' resulting in high cycle cost, *long-range aircraft are economical on long flights only*.

## 12.5 Fleet commonality

Aircraft operators are confronted with various aircraft type related costs. The first is the cost for training of flight crews and maintenance engineers working with that aircraft type. Second is the cost for maintenance itself, investing in spare parts and tooling and test equipment for a specific aircraft type. Every individual aircraft type generates its type specific costs. The more various aircraft types the operator has in its fleet, the more type specific cost this will generate. There is thus all the more reason for an aircraft operator to limit the number of different aircraft types in their fleet and seek for fleet commonality.

Fleet commonality is made possible by the aircraft suppliers by means of offering family aircraft types, aircraft of a single type coming in various capacities. This enables the operator to vary the capacity offered in the market within a single aircraft type. Prominent examples here are the A320 family, coming in three capacity variations, or the Boeing 737, offering even four variants.

These aircraft types offer engine commonality, an important element for lowering the maintenance cost, as the engine is the largest cost driver for aircraft maintenance. And with common engines, all components in the aircraft are identical as well.

Family types also offer cockpit commonality, allowing the pilot with a single type rating to fly on any of the various capacity variants. With this the aircraft operator can optimize its cockpit crew planning.

## Suggestions for further reading

Paul Clark: *Buying the big jets*. Routledge, 2006.

Paul E. Eden: *Civil aircraft today*. Amber books, 2008.

CHAPTER 13

# Aircraft supply and MRO

Aviation operations are about the use of the aircraft, not how it is designed and produced. However, when working with aircraft, an aviation operations professional will be faced with the suppliers of the aircraft, all the reason to spend some paragraphs on how the supply side of aircraft is structured.

Although aviation operational professionals do not deal with aircraft maintenance themselves, maintenance requirements largely define the availability of the aircraft, so we will consider these requirements below.

## 13.1 Original equipment manufacturers

The suppliers of aircraft are referred to as the OEMs (original equipment manufacturers). As such we see OEMs of airframes, engines and components.

### 13.1.1 Airframe OEMs

Airframe suppliers design, produce, sell and support aircraft. As such they are well known by the public as everybody has heard from companies like Boeing and Airbus, the world's two largest suppliers of commercial aircraft. The others are Embraer and Sukhoi. Indeed, the aviation industry relies on only a handful of airframe suppliers, with the Chinese company COMAC planning to enter the arena in the near future.

These airframe suppliers have a different background. Boeing, founded in 1917 by William Boeing, has been the supplier of large long-range bombers in WW2, after which this company has gained an early dominant position in the commercial aircraft industry. By taking over other US airframers, with McDonnell Douglas as the final one, Boeing has become the sole US supplier of commercial aircraft.

Airbus, on the other hand, is the result of a constant process of mergers between European companies. First on a national scale and finally on a European scale, the roughly 30 aircraft builders that existed in Europe in the early 1950s have been amalgamated into a single company.

The Canadian company Bombardier is also a result of mergers and takeovers, by which various Canadian aircraft builders are now within the single company. It is interesting to note that Bombardier is the only aircraft builder that is also active in other industries, including with trains and metro vehicles. As such, it has decided to divest from commercial aviation by selling its main aircraft programme to Airbus.

It is extremely difficult for a newcomer to enter the aircraft manufacturing business. Therefore, it is a giant achievement for the Brazilian company Embraer to be part of the scene. Embraer started in the late 1970s by introducing small turbo-prop powered regional aircraft. Step by step this company has expanded its product portfolio and its customer base to where the company stands today as the main supplier of small/medium sized jet aircraft. In order to remain competitive, this company has entered a joint venture with Boeing.

The Russian company Sukhoi used to be a supplier of military aircraft. Their commercial airframe is Russian's first attempt to enter the Western aviation business. Chinese COMAC is developing aircraft primarily for the Chinese market, but will in due course enter the global aircraft market as well.

Designing aircraft is a process that requires time, money and highly educated engineers. Designing the right aircraft is very difficult. In order to be commercial successful, an aircraft type needs to appeal to the needs of a wide variety of airlines in all continents. Niche aircraft, designed for special needs, are seldom commercially successful.

Designing aircraft, and setting up the manufacturing layout to build aircraft at a large scale requires huge investments, to be earned back over a long period. This is why airframe suppliers are financially far from "normal" companies, and it is fair to say that airframe suppliers would not be able to survive without financial support from their governments.

In order to extend the product life-cycle of an airframe to its maximum, airframe suppliers update existing airframe into "new" models by applying new engines and components on an existing, perhaps slightly improved, airframe. Designing and certifying an all new aircraft type is only done if there is no existing airframe available.

Producing aircraft involves a huge workforce. As demand for aircraft fluctuates with upturns and downturns of the world economy, these workforces are also in constant movement. Airframe companies are in a constant process of either hiring (in the upturns) or firing (in the downturns).

Selling aircraft has its own challenges as aircraft are sold to airlines in all corners of the world. This effectively means that the suppliers need a giant sales force that indeed covers all corners of the world.

The same applies even more to supporting the aircraft that are delivered and in operation at airline customers. Airlines need intensive support on training, engineering and parts supply. Manufacturers run massive support organizations with global coverage and flawless parts logistics.

Finally, and certainly one of the most important things that airframe suppliers do, is integrating all parts and components that they buy from other suppliers into an aircraft. For this reason, airframe suppliers are often called the integrators, as they integrate engines and components that are designed and produced by others into the airframe that they designed themselves.

### 13.1.2 Engine OEMs

The engine is the heart and soul of an aircraft. The engines define the aircraft's operational capabilities and, to a large extent, its commercial viability. Aircraft are designed around an engine, and this notion places the engine suppliers in a *vital leading position* in the development of aircraft. Also, here, the entire aviation industry relies on only four engine suppliers: US-based GE and Pratt Whitney, UK-based Rolls-Royce and the French company Snecma.

Developing new technologies and designing engines incorporating these technologies takes years and involves huge investments. Engine suppliers truly need visions on the future of aviation; they need to assess which engines the world will need in the future, and mistakes here can be lethal for the engine supplier.

Engine suppliers have intimate relations with the airframe manufacturers. They share technology and design details as the end product – an engine-powered aircraft – needs to be the result of a flawless integration of the engines and the airframe. Also on a commercial level, the relationships are intimate. The engine supplier is financially responsible for its engines, so engine suppliers are by definition risk-sharing partners in aircraft programmes. Here we see two business models: an aircraft can be equipped with only one engine type, by which the engine is called *single-source*, or the aircraft can be equipped with different engines from different suppliers, the dual-or even *multi-source* business model. In all models, the engine supplier takes the commercial risk for its engine, an aircraft's main cost driver. Just like the airframe suppliers, the engine suppliers also avoid designing all new models, but instead develop new variants of existing engine models.

In the dual- or multi-source model, the engine supplier needs to sell its engine to the end customer. And as engines require maintenance, the engine suppliers need to have support organizations with global coverage and parts logistics, even more so than the airframe suppliers. Engine suppliers offer their customers a total package including the acquisition of the engine and the maintenance for the years to come, ensuring income on the maintenance of the engine, in the cut-throat competition among engine suppliers often their main profit stream.

It is interesting to note that engine suppliers sometimes form joint-ventures, various engine models are designed and produced by two engine suppliers on a 50/50 base. As such, the world's most successful engine, the CFM-56 powering the Boeing 737 and the Airbus A-320, is a joint venture between US supplier GE and the French company Snecma, with production facilities at both sides of the Atlantic Ocean. The alternative engine for A320, the V2500, also is the product of a joint venture between US-based P&W and British Rolls-Royce. And as both P&W and GE considered developing an engine for the big A380 to be too risky, they did it jointly, resulting in the Engine Alliance 3200 powering the A380 as alternative to the Rolls Royce Trent 900.

Most notably GE and Rolls Royce have found commercially attractive applications for their engines outside aviation as well. The engine core of some of their models have found their way into the maritime world by powering naval vessels; electrical power stations are also operated by turbine engines based on aircraft engines. As electrical power stations run 24/7, these applications have generated enormous turnovers for the engine programmes.

### 13.1.3 Component OEMs

Both the airframe suppliers and the engine suppliers integrate a wide variety of components into their end products. This wide variety of components comprises of a wide variety in sorts of products, a wide variety of technologies applied and a wide variety of business models. "The" component business does not exist. It is not a single business but a variety of complex and simple products; most of them expensive, and, an important discriminator; components that are *exclusively designed* for a specific aircraft type versus components that are applied in various or even *most aircraft types* in production. When exclusively designed for a specific aircraft model, the component supplier shares the commercial risk with the aircraft supplier. When applied in various aircraft types, the component supplier cares less about the commercial success of a specific aircraft type.

Very expensive, technologically complex and designed for a specific aircraft type is the *landing gear,* supplied by one of the only two suppliers that dominate this business: US-based BF Goodrich and the French supplier Messier-Dowty, which is part of the also French component conglomerate Safran.

Expensive and complex also, but applied on various or even about all aircraft types, is the *avionics.* This stands for the full set of computers, antennas and sensors that are responsible for communication, navigation and surveillance (CNS) of the aircraft in operation. Leading suppliers here are US-based Rockwell Collins, part of UTC, Honeywell, and the French company Thales. Their sophisticated electronic products are found in all aircraft. Avionics comes at high market prices and moderate production cost; it is a commercially attractive business. And based on sophisticated technology as well as supplier reputation, it is difficult for outsiders to enter this business.

Aircraft need power of various types: *hydraulic* power for landing gear retraction, wheel brakes and flight controls as well as electrical power for the wide variety of electrically driven systems. Both have their specific suppliers. This also applies to an aircraft's *pneumatics*; air supply is required to pressurize the aircraft's cabin in the air and supply oxygen at comfortable temperatures. An aircraft's ECS (environmental control system) comprises a multitude of components, most of them aircraft-type specific and provided by a variety of suppliers. On the ground this all is powered by the aircraft's APU (*auxiliary power unit*), another expensive and aircraft-type-specific component.

Aircraft *interiors* form a special mix of aircraft specific components like wall panels, interior lighting and overhead bins, in combination with non-type specific products like aircraft seats, IFE (inflight entertainment) and internet *connectivity.* Aircraft type specific, or, better put, aircraft hull specific, are the *galleys* in the aircraft.

Just like the airframe- and engine suppliers, the component suppliers also have global product support in place, providing their end customer – the airline operator – with repair and modification services as well as parts supply. Support centres of large component suppliers can be found all over the world.

## 13.2 Maintenance, repair and overhaul

This textbook is about aviation operations and aircraft MRO is primarily engineering related. However, aviation operations professionals need a basic understanding on what MRO is and what complexities its entails. The operations professional will not have to deal with the challenges inside the hangar, but need a basic understanding of these challenges in order to manage the deployment of an aircraft.

An aircraft is composed of millions of parts, of which a few thousands need maintenance and can be removed or modified. Aircraft need maintenance, repair and overhaul in order to remain *fit for operation*. Operators can do this themselves or have it done by others. In both cases, it needs to be done well for three reasons.

First, Aircraft Maintenance, Repair and Overhaul (MRO) is well defined by the authorities. An aircraft only holds its airworthiness as long as the MRO on that aircraft is performed as per the manuals that are issued by the manufacturer. The *continuous airworthiness approval* makes aircraft MRO an obligation to the operator. So, the first and most important reason for which operators must maintain their aircraft "by the book" is the *legal obligation* to do so.

But even without this obligation, operators have a sound second reason to maintain their aircraft as good as possible: the *operational integrity* of the operator. Aircraft need tight flight schedules in order to be profitable, and passengers consider on-time performance as an airline's prime quality factor. Proper MRO is a precondition for an airline's on-time performance, as it minimizes the risk of technical failures of the aircraft in operation.

On a longer term, airline operators and the owners of the aircraft see a third reason for MRO, the *market value* of the aircraft. Well-maintained aircraft possesses higher market values than aircraft that received less attention, and certainly operators that keep their aircraft in the fleet for many years can only do so because they maintain their aircraft very well.

### 13.2.1 Legal framework

Certification of maintenance, repair and overhaul of aircraft and its individual parts is very complex, as it involves here a combination of *three interrelated certification processes*: the certification of the *organization* that performs the MRO, the certification of the MRO processes on an individual *aircraft or its parts*, and the certification of the *engineers* that perform the MRO work. This rule-making is based on ICAO's Annex 1 for personnel licensing and on Annex 8 on the airworthiness of aircraft.

MRO companies that provide their services to airlines need a Repair Station certification. The airlines in return are obliged to ensure that their aircraft are maintained and repaired by certified MRO companies.

Certification of the MRO organization, the so-called MOA (Maintenance Organization Approval) is the subject of EASA Part 145 and FAA Part 145. It stipulates how the company should be *organized*, which individuals have formal *responsibilities* and how the *quality* of the work must be guaranteed. It stipulates how all relevant *manuals* should be administered and kept up to date. It sets standards for *tooling and equipment* used in the company, with special emphasis on *test benches* and its calibration. The company has all the above well described in its MOE (Maintenance Organization Exposition).

Every MRO organization having the overall responsibility for the MRO on a certain aircraft, like an airline's Technical Department or an integrated MRO provider, needs to be certified as a CAMO (Continuing Airworthiness Management Organization), in which all aspects of the organization are described in a CAME (Continuing Airworthiness Management Exposition). This is all described in EASA Part-M and FAA AFS-300.

Here also the actual maintenance and repair that is done on an individual aircraft or its parts needs to comply with the so-called *Continuing Airworthiness Requirements*. The idea behind this is that once the aircraft and its parts have obtained an Initial certificate of Airworthiness (as described in Section 11.1), it needs proper maintenance and repair to retain the status of continued airworthiness as set in the OTC. For this, every aircraft needs to have an approved AMP (*Aircraft Maintenance Program*).

The basis for all rule-making here are the Maintenance Manuals and the Repair Manuals of the aircraft and its parts. These manuals were certified with the OTC of the aircraft and its parts. The basis is that all maintenance and repairs should be performed *as per these manuals*. Rule-making here consists of guarantees and evidence, by means of data administration and control checks on the work floor, that all maintenance and repair is carried out in accordance with the manuals.

When an individual part fails, and is taken off the aircraft, it is declared "unserviceable", which means that for this specific part/serial number, certification is suspended. Only by proving that the part is repaired and tested in accordance with the manuals, its certification will be re-granted by issuing an *Authorized Release Certificate*, like FAA's 8130 tag. This certificate is the evidence that the part complies with the TSO (*Technical Standard Order*) of the agencies.

Maintenance procedures are described in the manuals in great detail. At this level of detail, it is not always possible for authorities to control nor prove that all rules are met. At such detailed levels an AMC (*Acceptable Means of Compliance*) is sufficient for proving airworthiness.

Engineers that perform all this work need to be certified as well. EASA Part 66 and FAA Part XX define precisely what a person needs to know and what experience he or she needs to have to become a certified mechanic. It also describes how examination needs to take place for providing evidence to the existence or up to date knowledge. Training- and education providers need to comply with EASA Part 147.

Every MRO organization has one or more *Quality Assurance* managers on its payroll. They have great authority as they are certified to be a QA manager. They have the challenging task of ensuring that the company complies with all detailed rules and regulation at the three interrelated levels.

### 13.2.2 Maintenance repair and overhaul

MRO consists of three separate but interrelated pillars: maintenance, repair and overhaul.

*Maintenance* is done to maintain the technical qualities of the aircraft. It is performed on a regular basis, implying that it is *predictable and plannable*. Manufacturers define the required maintenance on the basis of time-related check intervals. These time-related checks are known as the A/B and C check, and the largest, after many hours of flying, the famous D-check by which the entire aircraft is ripped into pieces. In today's operation, operators combine parts of the various checks for synchronizing the check intervals with the operational requirements of the airline. We call this *equalized* maintenance and should be based on rules that are defined by the Maintenance Steering Group (MSG). Maintenance can partly be performed with the aircraft in operation but for the larger checks the aircraft needs to be inside a hangar.

*Repair* is a different thing. Equipment can fail after which it needs to be repaired. This is in principle *unpredictable* and therefore unplannable. Repair of a failing item in the aircraft is seldom done on an aircraft in operation. If something in the aircraft fails during operation, the failing unit is replaced by a functioning equivalent, after which the failing unit is repaired in a repair shop. A unit that can be replaced on an aircraft in operation is called a *Line Replaceable Unit*. It means that after replacing such a unit, the aircraft does not need to go through a test programme, and can operate as soon as the new LRU is installed.

In order to avoid the failure of a unit inside an aircraft becoming a driver for delays, the industry has adopted statistic indicators as to the reliability of a unit. Therefore, everything on the aircraft has a specific *MTBF* (Mean Time Between Failure) or *MTBR* (Mean Time Between Repair). Another way in which the industry has tried to avoid delays is by dividing all units inside the aircraft either as a *hard-time* item, or an *on-condition* item. Hard-time items need to be replaced after a certain usage in terms of flight hours or cycles, whereas an on-condition item remains in the aircraft until it fails. As a rule of thumb one can say that all components that are crucial for flight safety are hard-time, whereas a secondary system can well operate on an on-condition basis.

This also implies that not every part that fails need to be replaced immediately. The aircraft manufacturer has composed a so-called MEL (Minimum Equipment List). The MEL is the document that specifies which part may be unserviceable under what conditions so the aircraft may be dispatched and fly. The requirements that need to be fulfilled to enable dispatch with failed components are called 'MEL Relief'. If a part or component is not listed in the MEL, and it is unserviceable, that leads to an AOG (Aircraft On Ground).

Not everything that fails can be repaired. All that fails or needs to be replaced without being repaired are called *consumables*. Parts that can be repaired are called *rotables*.

We just stated that failure of something is unpredictable. However, with today's technology on sensors and the rapid development of digital data management we see a big trend in *predictive maintenance*, where on the basis of actual performance data of a unit, its upcoming failure can be predicted. This allows a cost-effective replacement of such unit just before it will fail.

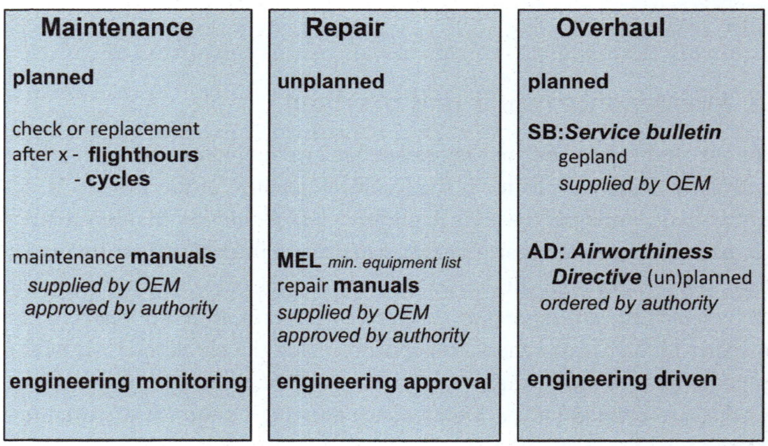


**Figure 13.1** Three pillars of MRO

*Overhaul* means totally disassembling a unit and rebuilding it with replacement parts for every part that needs replacement. After an overhaul the unit is like new; we call the status of a freshly overhauled part a "zero-hour" condition.

Included in the overhaul can be the *modification* of a unit. During the lifetime of a unit the manufacturer of that unit may well come up with improvements in the design or the application of better materials for parts in the unit. Such improvements are called *Service Bulletins* and are normally incorporated in a unit during an overhaul. If a component was ill-designed in the first place, the modification may well be declared obligatory by the certifying authorities, in which case the SB becomes an AD (Airworthiness Directive).

### 13.2.3 MRO intervals

MRO intervals are primarily defined in terms of flight hours – the aggregation of hours that the aircraft is flying. But for parts of the aircraft that become hot, like the engines, or parts that will stress, like the wing structure or the fuselage structure, we need to look not only at the flight hours but also at the cycles. A cycle is a flight, irrespective of the duration of that flight. A cycle involves starting up a cold engine to its operating temperature and back. A cycle also involves a landing, reason why MRO of the landing gear and the wheel brakes is defined in cycles. In order to keep MRO manageable, the cycles are translated into the flight hours, by applying a flight hour to cycle ratio. This ratio differs per operator, and depends on the average flying time per cycle.

### 13.2.4 Three levels of MRO

Once inside the hangar, an aircraft is divided into three different levels of MRO: airframe, engines and components.

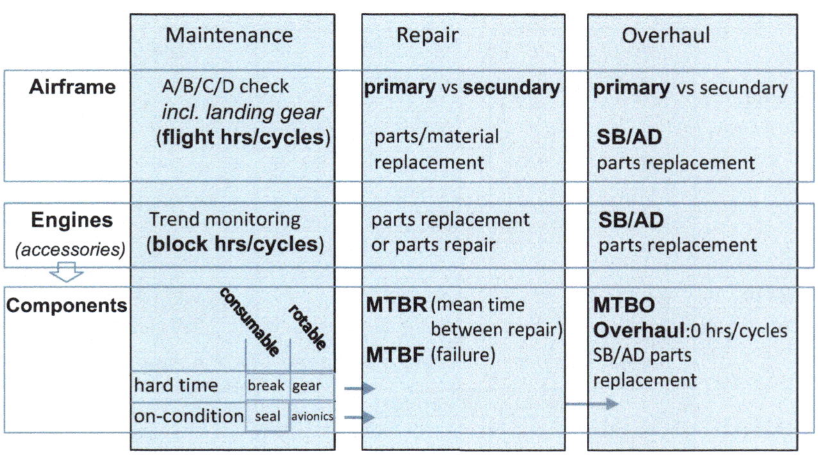


**Figure 13.2** Three levels of MRO

Airframe MRO consists of checks defined in number of flight hours. As such we know the A, B, C and the D-check. Operators can choose to combine various tasks in order to eliminate the need for one of the checks, or to ensure an optimal aircraft availability at peak demand. We call this equalized maintenance. When aircraft become older, mid-life, the airframe needs to be checked for material fatigue on a regular basis.

Engines do not need much maintenance while in service. The performance of the engine is closely monitored in terms of N2 (thrust) and TGT (temperature). When the engine performance drops below a certain minimum, the engine will be taken off from the aircraft to be fully revised in an engine shop. After this overhaul, the engine will perform as new.

Aircraft have many very different components installed. We divide them in hard-time components – parts that need to be replaced after a certain number of flight hours or cycles – and on-condition items – parts that remain in the aircraft until they fail. Hard-time components are parts of the aircraft that are vital for safe operation; on-condition components are elements that are nice to have in the aircraft, but they aren't vital for safety.

Another division between components is between the consumables – parts that are scrapped after use – and rotables – parts that are repaired or overhauled after use in order to be fit for use again after the repair.

Aircraft MRO is divided between Line maintenance versus Base maintenance. When the aircraft is in service it will need light checks and servicing like oil replenishment. This is done at the ramp during the TAP or at night, and all this is Line maintenance. A special responsibility for Line maintenance is repairing an aircraft with a technical failure in operation. Such aircraft is declared AOG (Aircraft On Ground). When the failing part is an LRU (Line Replaceable Unit), Line maintenance can repair the aircraft at the ramp. If the replacement part is not an LRU, the repair must be performed inside a hangar, often with a test programme to be performed after the repair.

Base maintenance is all that is done inside the hangar, with the aircraft not scheduled for operation.

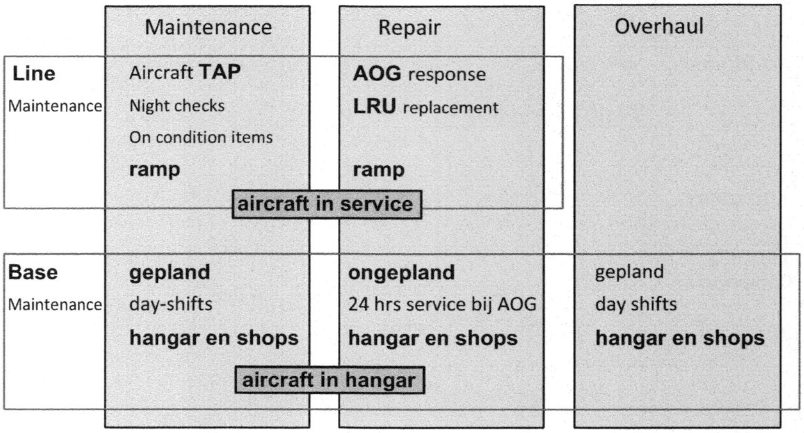


**Figure 13.3** Line versus base maintenance

## 13.3 Managing MRO

Parts logistics is, beyond obtaining and preserving capabilities, key in managing MRO. We just mentioned LRUs to be replaced at the ramp. This can only be performed when the LRU is available. Certainly, with technical failures at a destination airport, parts availability can be a challenge. But also at the operator's home base, or within base maintenance, availability of a component, or of piece parts for repairing a component, requires *sophisticated parts logistics*. An aircraft operator or its designated MRO provider can choose to hold an inventory of spare parts. The advantage is the immediate availability of the part at the location of that inventory. Setback is the capital cost of all these parts. For this reason, we see that more and more MRO providers also offer parts coverage by maintaining a parts inventory for many operators as well as spare parts pool arrangements, where members of the pool have access to a joint parts inventory.

With aircraft becoming more sophisticated and with the need for optimal aircraft availability, it becomes more difficult for operators to organize and manage all MRO by themselves. Only the very large MRO outfits can offer all MRO requirements in a cost-effective manner. For this reason, many operators outsource the more complicated MRO requirements to dedicated MRO providers. Day-to-day servicing, often up to B-check, is performed in-house by many, but all above the B-check, and certainly engine MRO and repair and overhaul of sophisticated equipment is outsourced to dedicated MRO providers. These can often also offer Line maintenance at destination airports.

## 13.4 MRO providers

In the past, airlines performed the maintenance, repair and overhaul of their aircraft themselves. Legacy airlines used to deploy large MRO departments with all required capabilities in-house. As of the 1980s, airlines – and certainly the new entrants – have decided to outsource part or even all of the MRO requirements of their fleets. Simultaneously, some large airlines have decided to turn their in-house MRO departments into separate profit centres, providing the MRO for airlines that have decided to outsource it.

This is why the largest players in MRO have their origin in an airline, or are still part of an airline company. MRO providers like LHT (Lufthansa Technik), AFI (Air France Industries/KLM E&M) or Delta Ops are the largest players. Even in case of bankruptcy of the airline, the MRO outfit can remain in place and be successful, like SRT (Swissair Technik).

MRO providers are primarily ranked by their *capabilities*. Aircraft MRO require a multitude of engineering capabilities in order to maintain, repair or modify aircraft, engines or components. They need quick access to repair parts. The more capabilities an MRO provider has at hand, the more services this MRO company can provide to its airline customer. As airlines outsourcing their MRO prefer to do this with a limited number or even a single provider, the MRO company with the most capabilities gets most of the business. And indeed, the largest company provides "nose-to-tail" services for an airline's entire fleet, with base maintenance, Line maintenance at all airline stations and network-wide spare parts supply covered in one contract, often against a dollar amount per flight hour produced.

This need for full capabilities has two effects that characterize the MRO sector. First, the winner takes it all. The providers with nose-to-tail capabilities are the major players as they manage the entire MRO spending of the aircraft under contract. Thus, we see a handful of large players with a multitude of small special providers below them.

Second, the need for access to all MRO capabilities makes that relations between the MRO providers can have different forms at the same time. On one hand they compete against each other, but as none of the players have 100% in-house capability, they also rely on each other for the coverage of MRO that they cannot perform themselves. So, besides being competitors, the large players are each other's customers or even partners as well. Furthermore, as no player in this world has a 100% in-house capability, commercial success of players is also defined by their capability to enter agreements with other players.

The MRO scene becomes even more complicated by the involvement of the OEMs. These OEMs defend their MRO business in order to earn a fair share of the MRO that need to be performed on their products. It is fair to say that over a 25-year operation, the initial value of the aircraft is again spent on maintaining that aircraft. Indeed, MRO is very big business in aviation. For the OEMs the main source for profit is in the sales of parts. Here we sometimes see intimate relations between OEMs and major MRO organizations, based on the agreement that the MRO organizations exclusively use OEM parts and OEM repairs. Such MRO organizations can become a partner in the OEM's global support outfit, with MRO outfits performing services on behalf of the OEM.

These OEMs aim for high market prices for their parts and repairs, while airlines need the MRO bill to be moderate. The gap between the two is filled by the so-called PMA providers (parts manufacturing approval), companies that produce a part that is comparable to the OEMs part and is certified as an alternative by the authorities. Repair shops can also be PMA providers by providing repairs for parts that the OEMs consider scrap parts. In the past, the tier-one airlines did not make use of PMA services, but as the quality of PMA services has improved and as PMA parts are sometimes delivered more quickly than OEM parts, the former perception of a lower quality of PMA has gradually disappeared (or is gradually disappearing). However, leasing companies often still regard PMAs as negatively affecting the value of their aircraft. They then still stipulate their aircraft's MRO to be on an OEM base. The fact that PMA does not imply lesser quality any more is perhaps best illustrated by the engine suppliers. In the cut-throat competition on engine MRO, the engine OEMs provide OEM services for the engines of their own brand and PMA services on engines produced by the competition.

Economies of scale are key to cost-effective MRO. MRO providers need quantity in order to economically justify investments in parts inventories, repair shops and test banks. Ensuring economies of scale becomes more and more challenging as technology developments are all aimed for improved reliability, increasing interval between checks and ease of maintenance.

MRO is complicated by the fact that the various parts of an aircraft require different forms of MRO. Some parts in an aircraft are predictable, like a wheel-break replacement after 1,500 landings. Others are unpredictable, like an avionics computer. Another complication is that all MRO need to be performed in accordance with complicated rules and regulations. MRO complication is further extended by the fact that aircraft fly over the world and nobody can predict when or where a part will fail.

One of the important developments in MRO therefore is the development of predictability. With aircraft being more and more equipped with sophisticated monitoring devices, MRO organizations obtain masses of real-time performance data. We now see the emergence of new service providers that specialize in defining reliability predictions on the basis of these data. This development is referred to as *predictive maintenance*.

Another important development coming from the complexities of MRO is the need for optimizing MRO operations. Here we see process optimizing programmes like LEAN and 6-Sigma more and more governing the MRO organizations. With optimizing the organizations comes the increasing sophistication of parts logistics, also key for cost effective MRO, and applying both to serviceable components required for an aircraft in service, as well as to piece parts in the repair shops.

## Suggestions for further reading

Harry A. Kinnison, Tariq Siddiqui: *Aviation maintenance management*. McGraw-Hill, 2012.

# PART IV
# Airlines

This part of this textbook is about airlines, the core players in aviation. Aviation is about connecting cities and regions and airlines are the *operators* that actually *produce and sell* these connections. They sell tickets to passengers, operate the aircraft and occupy the airports. They bring the world of aviation to the *consumer*, and if they fail, they go bankrupt. In this part of our book we will describe the fundamentals of how they do all this.

We will start Chapter 14 with an overview of specific elements that characterize the airline business, the *management perspective*. In Chapter 15 we will look into airline *economics*, as understanding the economics of airlines is indispensable for understanding airline operations. In Chapter 16 we will concentrate on the fundamental elements of airline strategic *planning* on fleet, network and scheduling. In Chapter 17 we will look at the day-to-day operational aspects of airlines, and the specifics of cargo operations.

CHAPTER 14

# Airline management

## 14.1 Legal framework

If a company wants to fly aircraft commercially, that company needs an *AOC*. FAA calls it the *Air Carrier Operating Certificate*; in EASA it stands for *Air Operator Certificate*. It means the same and this AOC does two things: it gives evidence that the holder of that AOC *complies to all rules and regulations*, and it defines the *nationality* of the holder flying under that AOC. An AOC is issued by the National Aviation Agency of that nationality, based on ICAO rulings annexes 6, 7 and 8, and outlined in detail by FAA and EASA. Most countries follow their standards, be it sometimes with minor differences. In all cases, an airline is called a *carrier* in legal terms.

There is a fundamental difference between aircraft type certifications, as described in Chapter 9, and an airline's AOC; it is held not by a company but by a *natural person* within the airline, often the VP Flight Operations. The AOC-holder in the airline is called the *Accountable Manager*. And indeed, he or she is accountable in person. This personal accountability ensures maximum integrity and sense for responsibility within airlines. Where compliance with type certifications can be controlled by (test) data and reported occurrences, an airline cannot be controlled permanently in day-to-day operation. The system of personal accountability is meant to ensure that airlines will always comply to the ruling underlying their AOC. And there are many rules to comply with. Let us have a look at the various responsibilities and obligations for an AOC holder, an airline.

An airline needs, both financially and technically, to be *fit and willing* to operate. This ICAO formulation forms the basis of a national AOC.

The airline needs to have sufficient *financing* and cash available. An airline needs a minimum financial solidity, ensuring that it can fulfil its financial obligations.

*Liability insurances* towards passengers or third parties are in place to the extent that the airline can fulfil its obligations, as defined by relevant liability treaties.

The airline has on its payroll a *sufficient number of trained and experienced licensed pilots and mechanics*. The airline needs to have a specified minimum number on the payroll, so not on a freelance basis. This is needed to ensure that in preparing, planning and executing flights, all flight operational rules and regulations are followed.

The *fleet is airworthy*. The fleet is defined in the AOC, and all individual aircraft, identified by their registration, possess a valid certification of airworthiness as well as an approved maintenance programme.

The airline management is *proven capable and well organized*. The AOC stipulates minimum requirements as to the organization of the airline, including appointment of the accountable manager(s) within that organization.

At the airline's destinations, *ground handling facilities* are all arranged for by the airline.

An AOC-holding airline needs to have an integrated *SMS* (*Safety Management System*) in place. An SMS is meant to optimize the safety of operation of the airline. Based on identifying potential hazards in the operation, procedures are described and enforced around these potentials. The airline needs to have a competent Director of Safety on its payroll, a function with far reaching powers within the airline organization.

An AOC also stipulates *national ownership* of the airline. Under FAA ruling, 75% of a US-based airline needs to be owned by US entities; the European Commission let EASA set the AOC standard at 51% European ownership. Most countries in the world follow the same standard. A single nationality of airlines is necessary as the system of air service agreements between states, as we discussed in Chapter 2, is based on nationality of the airline.

Also, interesting to know is that an AOC can be transferred from one airline to another. When two airlines merge, one of the two AOCs is no longer required for the new joint airline. Or, when an airline goes bankrupt, its AOC can form the legal basis for a new airline. Airlines like Swiss and Brussels Airlines operate under the OACs of their ill-fated predecessors Swissair and Sabena.

The AOC does contain a requirement for engineers on the payroll in the airline, but it does not contain rule-making on maintenance of the fleet. When an airline maintains its own fleet, it needs next to the AOC a Repair Station certification. Also, MRO companies that provide their services to airlines need such certification. Airlines in return are obliged to ensure that their aircraft are maintained and repaired by certified MRO companies. The legal framework for the maintenance of the fleet, whether it is done by the airline itself – *in-house* – or performed by a third-party MRO provider – *outsourced* – is described in Chapter 13 of this book.

### 14.1.1 Rules and regulations on flight crews

Flight crews need an airworthiness certification in person, based on annex 1 of ICAO. Captains need an ATPL (*Air Transport Pilot License*) and a *Type Rating*, First Officers need a *Civil Pilot License* with Instrument Rating and a Type rating, and cabin staff need a *Certificate of Demonstrated Proficiency* by the FAA or a *Cabin Crew Attestation* under EASA.

The Type rating is specific. A commercial pilot can be certified to one cockpit type only. Cockpit crew need to possess recent experience. The pilot's logbook, as attached to the ATPL, records the total flying hours. For actual operation, a minimum experience balance between the two cockpit crew members, defined in flying hours on the actual aircraft type, is stipulated by the authorities.

A cockpit crew member needs to be current, and is therefore subject to regular recurrent training. Pilot recurrent training is performed in the above-mentioned simulator as well. The final form of re-examination and inspection a pilot is subject to is the annual 'line-inspection' during which a pilot's functioning and performance is assessed by a pilot-inspector or pilot-examiner who observes the crew from the third (or observer-) seat in the cockpit.

In addition to their certificate, cabin crew need proof of their being trained on a specific aircraft type. Cabin crew members are allowed to fly on as many aircraft types as the airline wants as long as this is regulated in the AOC, e.g. many US registered airlines let their cabin crew operate on all types in the fleet, whereas many European airlines are restricted to four or five different aircraft types.

Crews are composed of cockpit crew and cabin crew, whereby the minimum number of cabin crew members is strictly defined by regulators, based on the number of Type 1 exits of the aircraft type/version. A rule-of-thumb is one cabin crew member per 50 passengers as the minimum.

And when on long-haul flights, a crew's operation is limited by maximum duty periods.

In the previous paragraph's, we saw that airlines and MRO companies need to prove that they are well organized and well structured. As we now discuss the licensing of individual persons, they need, as person, to prove fitness for operation by *periodical medical checks.* This applies to pilots, cabin attendants and also to Air Traffic Controllers. As airline regulation is incident-driven, periodical *psychological* checks have become mandatory for cockpit crew as well.

## 14.2 Business models

Airlines need to make profits in order to survive, and profits can be generated in various ways, resulting in a multitude of different airlines business models. We will go through them one by one, but we keep in mind that many airlines have become *hybrids* by adopting elements of different basic business models. We will only briefly define them here; differences in operations and network strategies will be dealt with in the later chapters of Part IV.

In air law airlines are called *carriers.* This phrase is applied to the naming of the various business models as well. So, in this section, we will see all the various forms of carriers.

### 14.2.1 The full-service carrier FSC

The main international carriers, often legacy flag carriers, are all called full-service carriers. They operate short- to medium- and long-haul routes with a fleet of mixed aircraft capacities, and all connect their flights at the home airport(s), the hub. That is why we can also call them the hub carriers. These airlines offer many connections, either non-stop or with a transfer at the hub. They offer various service levels, booking classes ranging from basic economy up to business class and sometimes even first class.

Most hub carriers connect their regional network with their long-haul network, and, depending on the location of that home airport, many connect their long-haul flights as well. Many of these FSCs have existed for a long period; we call them *legacy carriers.* These airlines used to be, or still are, *flag carriers* – the main airline of the nation of origin.

Most full-service carriers also transport freight: cargo and mail. They can do that in the bellies of passenger aircraft, or by dedicated freighter aircraft.

### 14.2.2 Low-cost carriers

These are the airlines that offer simple and standardized connections, often in only one booking class and mostly only short or medium haul. We mainly see these LCCs operating on a continental scale. Operating a *multi-base* network, they offer point-to-point connections only. LCCs do not connect their flights at the base but operate with very short turnarounds at the airports instead. By *maximizing the utilization of fleet and crews*, LCCs spread their fixed cost over a high production volume, resulting in a low operational cost base. In Chapter 15.3 we explain how this works. Important here is that LCCs give this low-cost base to the passenger by offering *low fares*. LCC products are affordable to a large public, so we see a business model where *high volumes* come with *low margins*.

The ultimate form of the LCC model is the so-called ULCC, whereby fleet, network and product are entirely standardized, and cost has been brought down to a minimum, resulting in ultra-low fares.

The LCC model is also applied on long-haul operations now, by airlines often being a subsidiary of an FSC. Here we see that it is difficult to apply the LCC business model successfully. Maximizing the utilization of fleet and crews more than the FSC, resulting in a lower cost base than an FSC, is often not achievable as FSCs generate high utilizations of long-haul aircraft as well, and crew efficiency on long-haul operations is predominantly defined by rules and regulations on rest periods between long-haul flights.

### 14.2.3 Regional carriers

These carriers concentrate on short- to medium-haul routes to regional destinations, often operated with small equipment. Many of these carriers have code-share agreements with FSCs or even operate as franchisee for an FSC. Here we see the regional carrier performing *feeder flights* to the hub airport of an FSC.

Regional carriers can have an important role in the country where they reside, certainly when aviation is the only practical mode of transport. Here we see regional carriers often operating on behalf of the government, or being subsidized on a per-route basis.

### 14.2.4 Leisure carriers

In the past, the typical leisure carriers were short- or medium-haul IT (Inclusive Tour) operators, and the IT operator sold its entire capacity to a tour operator via holiday packages. This has changed to a model by which the tour operator owns the airline as part of its holiday package production outfit. These leisure carriers are now increasingly operating to long-haul leisure destinations as well. Other former IT operators are transformed into LCCs. Presently we see some of these LCCs expanding to long-haul operations.

### 14.2.5 Freight carriers

Freight carriers do not operate passengers, but cargo or mail. They come in various forms. Many FSCs also carry operate freighter aircraft, while others are concentrating on freight only. They collaborate with freight forwards and cargo handlers. *Parcel operators* have the entire chain in-house, and operate large fleets of aircraft in hub networks. Freight consists of about 10% of total airlines turnover, and nearly 50% of this is performed by dedicated freighter aircraft.

## 14.3 The marketing challenge

With highly productive and cost-effective aircraft, operated in a liberalized environment, the aviation sector has become a *market-driven business* like other consumer businesses. This implies that, as in other commercial sectors, airline success is first and foremost based on marketing success. Obvious as it may seem, many (legacy) airlines had or still have difficulties in transforming their highly technical supply-driven organization into a demand-driven service organization. And indeed, airline success is primarily marketing success, resulting in passengers spending their money on buying tickets at the airlines. Selling tickets is the most important task for an airline and doing this at the highest attainable fares requires a profound understanding of marketing.

Paramount here is the notion of *perceived customer value* of an airline. Key here is understanding who the customer is, what that customer *needs or wants* and – on the basis how much that customer values the airline and its products – how much the customer is *prepared to pay*.

Resulting from the success of the LCC business model, many believe that all customers mainly want the lowest price. Looking at other consumer industries, like automobiles or smartphones, we see that this is not true: most consumers make a trade-off between price and perceived quality, and are prepared to *pay for that quality*.

The main element that drives perceived customer value is in *brand recognition*; the brand itself, the name of the company, is the basis for the perceived value. One can think of brands like BMW or Apple. The same applies to the airline business, where airlines like Singapore Airlines, British Airways or Delta successfully execute the art of *brand management*, indeed one of the most crucial departments in an airline organization. The essence of branding is that customers pay more for the brand, brand companies have higher incomes and consequently higher profits, by selling their products against higher prices than the less valued providers of objectively offering the *same or comparable* product or service.

Branding requires a successful execution of two steps: *recognition* by the market, and *appreciation* by the market. Only after this, brand recognition may appear. Market recognition is not easy for airlines. Airlines may be known in their home countries, but this is seldom the case in other countries. Hub airlines offering connections between two foreign countries with a transfer in the home country have an even greater challenge in being recognized as an airline offering a connection. Airlines need intensive market communication to become known in their target markets.

## 14.4 Customer management

Successful branding requires a clear definition of target markets, effective communication with these markets and fulfilling expectations by producing products of high quality, resulting in *customer satisfaction*. The latter is fundamental for the airline business as most money is earned from *repeat* customers. Customers prepared to pay more also fly more. Successful brand airlines have large first- or business classes and earn the bulk of their income on repeating long-haul premium passengers.

In the previous section, we introduced the notion of perceived customer value as the basis for market success, and the core of that is in the word perceived, which is definitely *not objective*. Objectively a VW- or Huawei product may be comparable or even better than BMW or Apple. But the customer perceives a value on the basis of fulfilment of needs, wants and expectations, which becomes complicated for airlines. Airlines do not produce hardware products; they produce connections whereby the customer is part of that production. Airline business is therefore a *service industry*, in which a passenger is a customer, and is judged by how far it fulfils his or her needs and wants, or exceeds expectations – all this at the moment of production: the airline transporting the passenger.

### 14.4.1 Digitization and social media

Airlines need to know and understand their passengers, and need to communicate with these customers. Here *digitization* enters the equation; airlines can communicate one-on-one with their customers on the basis of their customer knowledge and by using *social media* devices. Certainly, when it comes to disruption management, with delayed or cancelled flights and missed connections, the quality of customer communication is key to not only avoiding or minimizing the damage, but even *enforcing* the perceived value by exceeding expectations. Here we see airlines as quick adaptors of social media for two strong reasons: communicating with *passengers* on a per flight basis for disruption can be executed quickly and against low cost, and communicating with *customers* on remarks, suggestions and most notably complaints. Customers accept that their communication with the airline through social media is public, which helps the airline communicate more effectively. Other customers can see how complaints are treated positively, leading to a happy customer; this is therefore more effective than other costly marketing communications.

### 14.4.2 Frequent flyer programmes

Airline Frequent Flyer Programs (FFP) is an important tool for customer management, and is meant to create *customer loyalty*, a repeat passenger. Membership requires filling in a form with broad array of questions, resulting in the airline knowing that customer. By offering mileage and service perks the customer is stimulated to fly all his or her travels with that airline and indeed, recognized brand airlines have a large group of high-level membership customers, flying often and against high fares, generating a strong revenue stream for the brand airline.

FFPs are basically meant to let the regular customer collect millage-points to be spent on free tickets. This is only effective when the regular passenger sees true value in millage of that airline. Successful and effective FFPs let the repeat customer collect mileage on a broad network with many destinations.

## 14.5 People business

We all understand that airline business is technology and capital intensive. However, the airline business is foremost a people business. Airlines need *many staff* in order to operate the fleet, and as we saw that aviation is very much a service industry, we need staff interacting with passengers. Successful airlines understand that customer satisfaction can only be generated by satisfied staff. Airlines need staff being happy, proud and satisfied working for that airline. When employment satisfaction drops, customer satisfaction drops as well, resulting in low perceived value and ultimately lower revenues.

As said, airlines need many staff. In Section 15.6 we will look at crew planning, and we will see that an airline needs many pilots and cabin attendants to run the operation. Ground staff and engineers are needed, and often on a 24/7 basis. Keeping all these people happy, committed and motivated can be a true challenge to the airline's management, but it is a prerequisite for success.

When staff is unhappy, their attitude to the passenger will degrade. And when staff is so unhappy that they go on strike, the airline is in real trouble.

Also on the ground, at the home airport and in the hangars and offices, airlines need many staff. Airline business is very labour intensive, and the *cost base* for all that labour is the *country of the airline's AOC.* Airlines can only limitedly export their labour-intensive activities to low-wage countries in the way that other sectors of the economy do. Indeed, when looking at airline cost differences between countries and continents, labour cost base differences are the paramount basis for airline cost differences.

## 14.6 Airline ownership

In Section 14.1 we saw that an airline needs to have an AOC where the *nationality* of that airline is defined. This has far reaching consequences for how the airline industry is structured. Where in many industries the multinational business model has become the standard, airlines need to possess a *single nationality.* This effectively means that an airline needs to be owned for at least 51% in most countries, or even for 75% in the US, by entities holding the same nationality as the airline. The reason for this is the nature of the Air Transport Agreements that we discussed in Chapter 2. Airlines are allowed to operate under these agreements on the basis of their *nationality.* Thus, airline nationality is required for the execution of the ATA structure of the aviation industry.

Where many successful airlines seek to expand their business to other regions of the world, they find themselves limited in the possibilities to truly penetrate in these other regions. Airlines need a partner company in the country where they want to expand and have to accept that national partner to hold a 51% majority in such joint undertaking. The need for a single nationality by an airline is a complicating factor in the processes of airline cooperation and consolidation.

With the emergence of multilateral Air Transport Agreements, we see the ownership rules sometimes becoming wider as well.

## 14.7 Airline competition

The airline business is extremely competitive, where airlines compete for market share and for the premium customer. The reasons why this part of the aviation business is so competitive is twosome.

First, airlines often produce *comparable* or even identical products. They all fly aircraft of the same types to the same destinations, often even at the same time. This leaves the airlines with only two competitive tools: *brand recognition or fare level*. Indeed, airlines lacking an established brand recognition compete on fares, with margin dilution as a consequence. Brand airlines compete on (in-flight) quality levels and on uniqueness of *destinations* offered. Many airlines compete on *frequency* levels, offering more departure times to the market. But again, airlines produce comparable products, and once an airline introduces something new, be it a new destination or a larger seat, other airlines will follow suit. Indeed, airline competition is a rat race.

The second reason for competitiveness is in the observation that it is relatively simple to *start* a new airline. Many airlines with strong competitive positions are relatively young airlines. These new entrants may introduce new business models, as did the LCCs, or very successfully copy existing airlines, like the new carriers from the Gulf region. New airlines can establish brand recognition quickly, making the battle for brand recognition an ever ongoing one.

## 14.8 Airline cooperation

The strong competitive forces described above lead airlines to seek to cooperate with one another under the maxim "if you can't beat them, join them". By joining forces, airlines try to improve their competitive positions in the market. This cooperation comes in many forms, and we will describe them in this section. But not before we have observed that airlines *always* have cooperated where necessary. Competitors though they may be, airlines provide technical assistance to each other. Every airline has a technical outfit and a spares inventory at its home base. If a visiting airline has a technical problem, and the home carrier can help, they will do so. At important destination airports, airlines may cooperate on a joint spare inventory. If an airline has a cancelled flight, other airlines take care of the stranded passengers.

### 14.8.1 Code sharing

A common way of cooperation between airlines is in code sharing. Every airline has an airline code, and its flights are numbered with this code. When a customer goes to the website of an airline, he can purchase any flight operated by that airline, all with the flight code of that airline. Code sharing means that airlines offer a flight under their own airline code that is actually operated by another airline. This code share can be *single-sided*: an airline sells a flight under its airline code that is not operated by that airline at all. Code sharing can also be *two-sided*: both airlines operate the route, and both airlines offer these flights under their own airline codes. The single-sided code sharing is often used among long-haul airlines and regional airlines, whereby the long-haul airline sells a final destination under its airline code, and the final leg of the route is operated solely by the regional operator. Airlines like Alaska Airlines or Air Baltic operate in this manner. The advantage to the long-haul airline is that it can offer *more destinations* to the market; the advantage to the regional airline is that it will get more passengers.

The two-sided code sharing is often used between two long-haul airlines that can jointly offer *more frequencies* under their own airline code.

Danger of code sharing is when the cooperating airlines have *differences in service levels* or brand recognition. If a passenger buys a ticket at a quality airline, that passenger expects that quality throughout its entire journey as he or she bought that journey under the airline code of that brand airline. If the final leg is operated by a carrier of lower quality, this passenger may be dissatisfied. Therefore, for successful code sharing the partner airlines need comparable levels of quality.

### 14.8.2 Alliances

Successful code sharing needs a synchronization of schedules between the sharing airlines. Additionally, cooperating airlines have the need to standardize their product quality. Frequent flyer passenger wants to earn mileage on the partner flights as well. Here we have the base ingredients for airline alliances.

Hub airlines connect their regional system with their long-haul system at their home airport. They fly long-haul to destination airports. These destination airports are hub airports as well, with a home carrier connecting its flights. By synchronizing schedule times and standardizing product quality, alliance airlines connect their network system with the network of the partners. This ensures passenger feed for the long-haul flight at both the departure- and the destination airport. The result is an airline alliance that can bring a passenger from any departure airport to any arrival airport in the world.

Airlines within an alliance integrate their networks: on the long-haul flights between the hub airports they no longer compete; instead, they cooperate by code sharing and often even share the income from the route.

At their home airport, alliance members perform the Turn-Around Process (TAP) of the visiting alliance partner aircraft. This results in huge cost saving for alliance airlines as passenger, cargo and aircraft handling at destination airports is expensive. Home airport ground operations are mostly performed with fixed capacity and fixed costs, which can be spread over more TAPs, with more alliance partners operating to the home airport.

There are three alliances – the Star alliance, SkyTeam and OneWorld – of which most larger hub airlines are members. A few large hub airlines, like Emirates, are not members of an alliance. Most LCCs are not members either. These airlines decided to beat the competition instead of joining them. Looking at the three alliances we see combinations of a few very large airlines encircled by smaller airlines. This implies that most of the shared benefits of alliance operations will often have a financial reconciliation to cover the cost-benefit imbalances resulting from the size differences.

### 14.8.3 Joint ventures

Alliance cooperation on the large routes between the hub airports can become very intense to the extent that the cooperating airlines decide to jointly operate these routes. This means coordination of flight schedules, standardization of service levels, *sharing the income*, increasingly also sharing the *profits,* and sometimes even taking an *ownership share* in each other. The result, a joint venture, covers an entire market.

A joint venture implies a common network operation, an optimized joint fleet strategy and an integrated schedule. On the demand side, a joint venture implies a standardization of fare systems and service levels, reason why these airline joint ventures require an immunization for anti-trust legislation by their authorities.

On the financial side, we see two levels of integration: the first form is an equal revenue share, but we also see an equal cost share, which, in combination with equal revenue share, results in *profit* share. As the joint profitability depends on the joint cost, we see strong JV partners taking shares in the smaller ones in order to obtain managerial influence on the cost management of these smaller partners.

Although the majority of joint ventures take place within an airline alliance, we also see JVs between airlines outside alliances, albeit that often the smaller entity in such JV becomes member of the larger partner's alliance. Nevertheless, it is fair to say that contemporary strategies of the large airlines are more concentrating on joint ventures than on the alliance model, and the development of joint ventures will become the paramount factor in airline development and airline competition.

In Chapter 2 we discussed the Open-sky agreements between states, and the JV model operates, in general, under such liberal agreements.

### 14.8.4 Take-over

When, in a joint venture, one airline buys a stake in the other, it constitutes a partial take-over of that airline. In the joint venture between Delta and Virgin Atlantic, the former bought a 49% stake in the latter. Here we see the impact of what was discussed in Section 14.6 on airline ownership. Within a country or a continent like the EU, the ownership limitations do not exist. In the US, we saw the largest three airlines taking over the three next largest, and in Europe we see powerful airlines like Lufthansa, British Airways and Air France taking over smaller European airlines. This continuing process is called *consolidation*, with the expectation that gradually the airline business will consolidate into a handful of very large holding firms, owning a multitude of airline brands. It is however unknown when the strict nationality regulation will evaporate or disappear, and as long as the major states like the US or China hold firmly to their nationality regulations, airline consolidation will remain within the structures of the joint venture. The multinational single airline is still far away.

### 14.8.5 Franchising

Some smaller airlines do not operate under their own name. They fly on behalf of another airline, under its flight code and with aircraft painted in the livery of the other airline. This business model is used in the US, where franchise airlines operate the regional feeder routes for the mayors in and out of their hub-airports. In Europe, we see the Scandinavian operator Sun-Air operating a wide array of regional operations under franchise with British Airways. In all cases franchise is primarily done on regional routes, where the independent regional operator has the brand benefit, and often also a secured income base, of the franchiser that the franchisee airline hardly could do without.

## 14.9 Volatility of demand

Demand for air travel is very volatile and prone to unexpected disruptions. Airlines need strong cash positions in order to cope with the unexpected. Fortunately, these disruptions seldom change the competitive landscape as most disruptors hit all airlines equally.

First volatility of demand is in *changes in the economy*. When the world economy goes up, demand for air transport *grows faster*. But when the economy goes down, this demand *goes down faster* as well. The trick here is to gain from the upside optimally in order to be well equipped financially for the downturn that will ultimately come. In the economy, these ups and downs are called the cyclic nature of the economy. In aviation, we therefore speak about the *cyclic nature* of the aviation business.

But besides cyclic ups and downs in the economy, aviation demand is also volatile for events in the world.

*Political crises* may damage the airline business. Aircraft overfly many countries on the way to their long-haul destination, and a political crisis in one of these may result in the impossibility to overfly that country. The result may be a longer flight time, disrupting the airline's schedule. The result also may be that airlines cannot fly to certain destinations any longer, resulting in excess fleet capacity.

The *outbreak of lethal diseases*, like SARS or Ebola, has a direct effect on the airline operating in or to the area where the disease is.

*Nature disasters* can temporarily affect demand for air travel to and from the affected area.

## 14.10 IATA

IATA (*International Air Transport Association*) is an organization of airlines just as ACI is an organization of airports.

IATA was formed in 1945 by the leading airlines at that time, and provided input for the Chicago Convention. This has resulted in bonding between ICAO and IATA that still exists. IATAs headquarters is, just like ICAO's, located in Montreal, Canada. As an organization of airlines, one of IATA's tasks is to guide and assist airlines in complying with complex rule-making. IATA provides courses and training for airline staff meant to keep staff educated and up to date with changing rule-making. A formal role for IATA in applying rules and regulations by airlines is in IOSA (*IATA Operational Safety Audit*). This is a quality audit that IATA performs on airlines. The IOSA certification, issued after a successful audit, is legal evidence for an airline that it complies with all rules and regulations. The IOSA audit is even mandatory for airlines in many countries. IATA also performs auditing on travel agencies and cargo agents.

IATA has issued two-letter codes to all airlines. This code is found on every flight an airline operates. A flight number starts with the airline code.

IATA has issued standard abbreviations of airports in a three-letter code. Where flight departments use the ICAO four-letter abbreviations, passenger-related departments use the IATA code, which is printed on the luggage tab.

## Suggestions for further reading

Ahmed Abdelghany, Khaled Abdelghany: *Modeling applications in the airline industry.* Ashgate, 2009.

John Wensveen: *Air transportation, a management perspective.* Routledge, 2015.

Nawal Taneja: *Airline industry.* Routledge, 2016.

Stephen Shaw: *Airline marketing and management.* Routledge, 2007.

## Interesting websites

IATA.org

Oneworld.com

Skyteam.com

Staralliance.com

# CHAPTER 15

# Airline economics

Anyone working in aviation operations needs a clear understanding of airline economics. Of all the different parties involved in aviation, airlines are, unlike NAAs, ATM organizations or airports, the ones that *need to make profits* by operating aircraft and transporting passengers or cargo. Airlines need to make profits in order to survive; a consistently non-profitable airline will go bankrupt. And indeed, many airlines disappeared due to their inability to be profitable. A state or a private owner may well decide to support its airline by compensating losses that the airline generates, but due to the cost of aviation today, this is difficult to maintain in the long run.

The way in which airlines operate is defined by *economics*; we need to understand the basics of airline economics in order to understand the way airlines operate. Airline economics is a challenging element of airline operations as these operations come with enormous amounts of money, both on the cost side and hopefully also on the revenue side. Here we see that the margins of the airline business are narrow as the difference between profit and loss is small. There is thus all the reason to spend an entire chapter on airline economics.

## 15.1 Key performance indicators

Airline economics is about cost and revenues. Before we can look at these, we first need to understand what airlines actually produce and understand the *measuring units* for that production. Cost levels can only be understood if we know what was produced against the costs. Furthermore, the fixed cost of an airline needs to be *divided by the production* in order to calculate the fixed cost per production unit. So, let us take a closer look at production and revenue KPIs.

### 15.1.1 Available seat kilometre

Airlines are often ranked by the number of aircraft they operate. And if we know how many seats these aircraft have installed, we know the number of seats that the airline deploys. But as airlines produce connections, we need to know *how far* these seats were transported in order to rank them on the basis of their production output. We need to add *distance* to that seat.

This is exactly what the ASK (available seat kilometre) tells us; it is a KPI that includes the *seats and the distance flown* in one figure. The ASK is simple to calculate. If an aircraft with 150 seats flies a distance of 1,000 kilometres, it has produced 150*1000=150,000 ASKs. This simple example also tells us that ASKs come in huge figures; large airlines produce ASKs in the billions on an annual basis. In many countries distances are not measured in kilometres but in (nautical) miles. Here, airlines produce ASMs (available seat *miles*), but the calculation remains the same. The 150-seat aircraft in our example above flies 1,000 kilometres, which equals 539,96 nm, producing 80,994 ASMs.

The ASK or ASM is the KPI that tells us the *exact quantity of airline production*. It is obvious that airlines operating large aircraft over long distances produce more ASKs than regional airlines operating small aircraft over short distances.

For cargo transportation, we do not use seats, but tonnes. Here we use the available tonne- kilometre, the ATK or ATM.

### 15.1.2 Revenue passenger kilometre

Airlines are often rated according to the number of passengers they carried. But airlines transport these passengers over a certain distance, so if we want to know what an airline actually did with all its passengers, we need to know not only how many passengers were transported but also *how far* these passengers were transported. Again, we need distance in the equation.

This is the RPK, the revenue passenger kilometre. It tells us exactly not only how many passengers the airline transported, but also how far. And the RPK also is fairly simple to calculate. If an airline flies 130 passengers over a distance of 1,000 kilometres, it produced 130*1000=130,000 RPKs or – in Nautical Miles – 70,195 RPMs.

Here also, if we look at the production of a cargo airline, we take the tons that were sold – the RTKs.

### 15.1.3 Load factor

In the example above we see a relation between the ASK and the RPK. We took as example the same aircraft flying 1,000 kilometres, with 150 seats installed and 130 seats filled by passengers. By dividing the RPKs by the ASKs, we see the percentage of ASKs that were actually sold. By dividing 130,000 by 150,000 we get 86.6%; this is the Load Factor for that flight. When an airline divides all the RPKs for a year by the ASKs produced in the same year, it calculates its *average load factor* for that year. Airlines also call it their *system load factor*. This is a fundamental KPI revealing how much of the airline's production is actually sold.

Airlines also look at the load factor on a per flight basis, a KPI relevant for commercial- and capacity planning. An individual flight should preferably reach its *break-even load factor* – the percentage of the capacity sold at which all costs for the flight are covered. On the other hand, the capacity sold should not exceed the *planning load factor*, above which the aircraft deployed is considered too small for the demand on that route. When the aircraft is too small, *spill* will occur – passengers that wanted to fly with the airline but went to the competition.

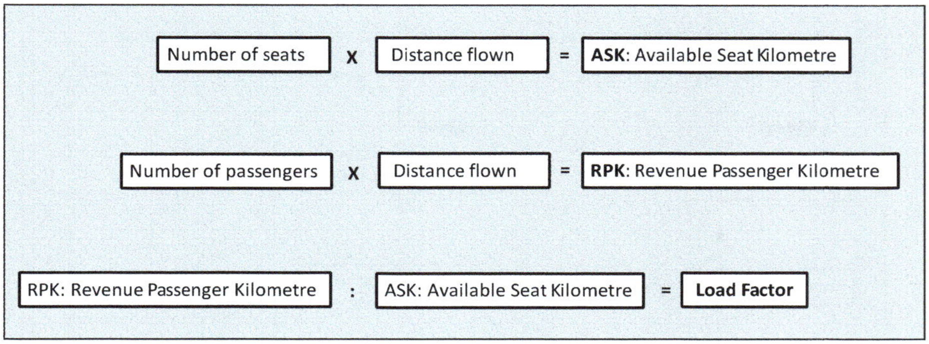


**Figure 15.1** Airline production KPIs

### 15.1.4 Cost per ASK

When we take the total cost of an airline per year and divide it by the total ASK production over that year, we get the cost per ASK (CASK). This is one of the most important KPIs; we call it the *unit cost* or cost per unit. This is the KPI by which airlines can benchmark each other on cost efficiency, and it is the CASK that tells us how cost-efficient an airline actually is.

### 15.1.5 Revenue per ASK

When we take all revenues of an airline by selling tickets, say per year, and we divide that by the total ASK production over that year, we have the Revenue per ASK (RASK). This KPI tells us how the products of an airline are perceived by the markets, where quality airlines have higher RASKs than airlines with lower perceived values. It also shows how relevant the connections that an airline produces are. The RASK is also referred to as the airline's *yield*, the income per unit produced.

Airlines often publish their average revenue per passenger. This is indeed an interesting KPI that tells the airline how much on average a passenger spends with that airline. The problem here is that this figure does not include the unsold seats. The importance of the RASK is that by dividing the revenues by the *total production*, the dilution of income as a result of flying empty seats is included in this KPI.

### 15.1.6 Aircraft utilization

Commercial enterprises need a maximum utilization of their capital goods in order to optimize profitability. Airlines are no exception to this; the *total number of flight hours per aircraft* or per fleet of aircraft of the same type, per year, is a crucial KPI for measuring an airline's cost efficiency. When this utilization is high, the cost of the capital good – the aircraft – can be spread over more production, lowering *the cost per production unit*. The ASK itself here is insufficient as this KPI does not tell us how many aircraft actually produced all these ASKs.

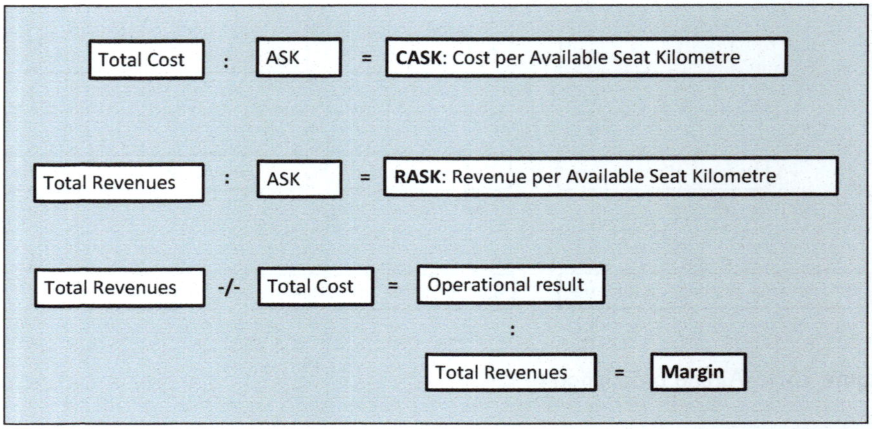


**Figure 15.2** Airline economic KPIs

### 15.1.7 Margin

One of the most important financial KPIs in business, and in the airline business, is the margin. Take all revenues and subtract all cost. This results in the operational result. By dividing this by the total revenue, we have the margin: *a percentage of the revenue that remains for profit* after deduction of cost.

As we just subtracted *all* cost, we calculated the *gross* margin. If we take all revenues and subtract the *variable cost*, and then divide this by the total revenues, we get the *contribution margin* – the percentage by which the *fixed costs are covered* by the revenues. This also is an important financial KPI for airlines, as it tells the airline management against which fare levels the fixed costs are covered. This is important to airlines, as they need to fill their aircraft at low demand also. Low fares are the best way to persuade passengers to fly.

Inability to cover the fixed costs by the fare levels offered in low-demand periods is a prime cause for airline bankruptcy.

## 15.2 Different types of costs

Airline costs come in different varieties, such as fixed cost versus variable cost and direct cost versus indirect cost. We need to understand the differences between the various costs. Cost competitiveness between airlines is foremost found in the fixed cost, whereas variable costs are more or less identical among all airlines.

### 15.2.1 Fixed cost

Fixed costs are those that are *constant during a defined period*. If an airline buys an aircraft and depreciates it over, say, 15 years, this cost will be equal each year during that period. If the airline employs a cabin attendant on a one-year contract, that cost will be fixed for a year. Here we have the largest fixed costs for an airline, its fleet and its staff. And if we divide the fixed cost by the ASK production, we come to the fixed cost per unit. If we divide the fixed cost for the aircraft by its hourly utilization, we have the fixed cost per flying hour.

### 15.2.2 Variable cost

Variable costs are costs that come with production. If an airline flies an aircraft, the *costs that come with that flying* are variable. The more the airline flies, the higher the variable cost become. Fuel cost is the best example. If we don't fly we don't have fuel cost. If an airline produces many flight hours, it will burn fuel accordingly. It is important to understand that most, if not all, variable costs are *about equal* for all airlines. Variable costs are also known as *expenses*, as the variable costs consist of bills to be paid.

Variable cost come in various forms of variety: we have cost per *flight*, cost per *flight hour* and cost per *passenger*. Fuel and navigation costs vary with flight *distance*.

### 15.2.3 Direct cost

Direct costs are all costs that are *directly linked to the production*. Nearly all variable costs are direct costs. In general business administration depreciation is regarded as an indirect cost, but airlines consider the ownership cost of the fleet as a direct fixed cost as it can be directly divided by the production.

### 15.2.4 Indirect cost

All costs that *cannot be linked to the production* are indirect cost. The costs for the head office and the managers are indirect costs. The older legacy airlines, especially, often have high indirect costs. Indirect costs are unavoidable, but the actual level of indirect cost is a matter of organizational efficiency. Indirect cost, like fixed costs, needs to be divided by the total

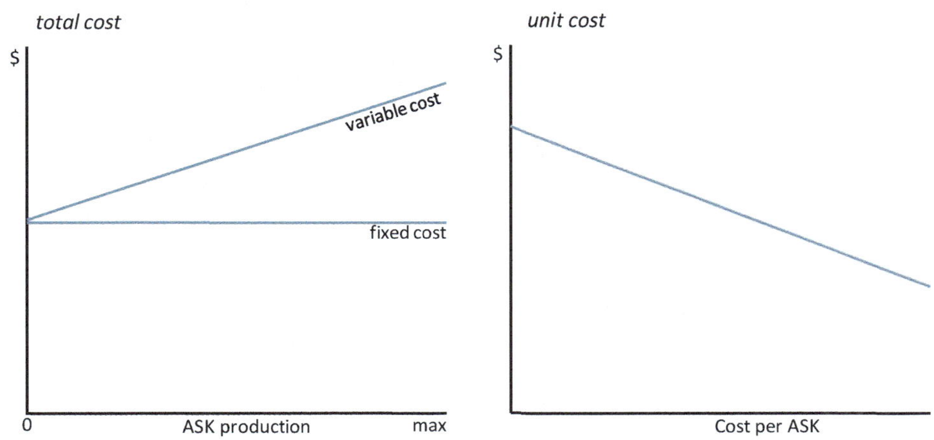


**Figure 15.3** Unit goes down with increasing production

production in order to arrive at the indirect cost per production unit. The indirect part of the unit cost says a lot about airline organizational efficiency. Here also, the more the airline produces, the lower the indirect cost per production unit. This is where the quest for scale comes from: the more an airline can produce with the same indirect cost, the more cost efficient that airline becomes.

### 15.2.5 Marginal cost

Marginal costs are costs incurred for producing an additional unit. For airline economics, this cost type is of limited value. Variable costs are incurred when an aircraft is in operation, whereby the total cost is only slightly dependent on the number of seats sold. Some variable costs are paid per passenger; these costs can be considered as marginal cost.

## 15.3 Direct operating cost

Now that we understand the different types of cost, we can start looking at the cost structure of an airline. For operating aircraft, we need to divide the fixed cost by the total production or utilization, and we have to add the variable cost on a per flight basis. Indirect costs also need to be divided by the total production. Adding these up we can calculate the total cost of the airline. Dividing that by its total production gives us the airline's *unit cost*, the CASK. We can also calculate the total operating cost (TOC) on a per flight basis.

Divided by the number of seats in the aircraft, we can calculate the TOC per seat. These calculations are needed for calculating the profitability of an individual route, or comparing different aircraft types on the same route, as we will see in Chapter 16.

In Chapter 16, we will look at the Direct Operating Cost (DOC): all costs, fixed and variable, that can be *directly linked to the production* of an airline.

### 15.3.1 Ownership cost

An airline needs to have a fleet of aircraft at its disposal in order to fly passengers or cargo to its destination. This airline can buy that aircraft or lease it on a financial or operational lease, as we discussed in Section 3.3. The decision will be made on the basis of the airline's financial capability to buy, or an airline's required flexibility as we will discuss in the next chapter. The actual ownership cost for the aircraft may not differ much.

In case of a buy, the airline has the cost of depreciation. This is not a real cost but an activity in the books. The real cost is the amortization of the loan needed to finance the aircraft.

Based on an amortization period in terms of years and the assumed remaining value of the aircraft, we can calculate the depreciation cost. We need to add the interest paid on the loan to derive at the actual *fixed* ownership cost per year for the aircraft.

In case of a lease, it is simpler, as we take the monthly lease rental. Multiplied by 12 gives it the annual *fixed* ownership cost for that aircraft.

In all cases, we need to add the hull insurance to the outcome. Typically, this is about 0.05% of the catalogue value of the aircraft. For insuring a used aircraft, we would take the actual value of that aircraft. Hull insurance is not an obligation, so larger aircraft may well build up a reserve on their balance sheet instead.

Ownership costs need to be paid in hard currencies, mostly in US dollars.

### 15.3.2 Fixed crew cost

An aircraft needs a full crew to fly. It needs a cockpit crew, consisting of two pilots, and a cabin crew whose quantity depends on the number of seats in the aircraft. These crew members are staff and on the payroll of the airline for a given period. The aircraft needs to fly many hours per day, much more than a human can work in that day. Thus, the airline needs *many crews per aircraft*. Members of the staff generate significant costs that include *payroll, social security, pension and insurance*, all in compliance with the labour legislation where the airline resides or where the airline has that staff on its payroll. These costs need to be paid in the currencies of that country.

An airline also needs to ensure that its crew members are current, so *training costs* forms another part of the total crew costs, to be paid in hard currencies.

This total cost for employing and deploying crews is a huge *fixed* cost. Flight crews are partly aircraft-type specific. Crews are assigned to a certain aircraft fleet. If we divide total annual crew cost for a certain fleet of aircraft by the total annual aircraft utilization of that fleet, we have the crew cost per flight hour for that fleet.

### 15.3.3 Maintenance cost

Aircraft need maintenance as we saw in Chapter 10. Calculating the maintenance cost is not easy since maintenance costs are *a combination of fixed cost and variable cost*, depending on the level of in-house versus outsourced maintenance.

In-house MRO comes with fixed cost for hangars, facilities, tooling and equipment, and staff on the payroll. All that the MRO organization buys, like spares and consumables, are variable costs. Costs for outsourcing repairs or services are variable. When, at the end of the year all costs are *added up and divided by the utilization of the aircraft*, one can calculate the maintenance cost per flight hour per aircraft. The fixed maintenance costs are in local currencies, the variable costs in hard currencies.

When maintenance is outsourced altogether, in a nose-to-tail service arrangement against a cost per flight hour, these *maintenance cost become variable*.

### 15.3.4 Fuel cost

Fuel is a *variable cost* by definition; the airline burns fuel when flying the aircraft. This fuel needs to be bought and paid. This is why we say that fuel is the *largest expense in an airline*. Calculating fuel cost is simple: by multiplying the amount of fuel consumed with the price of the fuel we derive at the fuel cost. When calculating actual fuel cost we need to know the actual fuel price at the departure- and destination airports; these can differ profoundly, as we saw in Section 3.3. At a given airport all airlines, in principle, pay the same for fuel. Airline managers are grateful for this as it means that fuel price is not an aspect of airline competition. This is vital for the largest airline expense, which is entirely unpredictable over time and must be paid either in dollars or euros.

It is indeed in principle that the same price is applied to all at a certain airport. In reality actual prices may *differ* on the basis of contracted volumes; fuel hedging; or, as is the case with Delta, an airline owning a refinery by itself. Fuel *hedging*, an agreement on a fixed fuel price over a certain period, can be very beneficial if the fuel price rises during the contracted period. Hedging however can be very dangerous if the actual fuel price drops, leaving the airline to pay significantly more for fuel than its competitors.

### 15.3.5 Ground handling cost

Regarding the ground handling cost, all ground cost related to the TAP (Turn Around Process) of the aircraft – a *mix of fixed and variable cost* – is applicable here also. Many airlines have their own ground handling at the home airport(s) where fixed costs for staff and equipment, mostly in local currencies, are generated. Dividing this by the total TAPs performed in a period, we can calculate a cost per turnaround. This cost is considered as *variable on a per flight* basis. When the TAP is outsourced at a destination airport, these costs are truly variable on a per flight basis, and need to be paid in hard currencies.

### 15.3.6 Airport charges

Airports charge the airline using that airport a service charge per passenger, covering the cost of the airport landside infrastructure. At many or all airports, a security charge is added to this, and in many countries an airport tax is charged as well. The total is *a variable cost per passenger*; this is not only a variable cost but a marginal cost as well.

### 15.3.7 Landing fee

Airports charge the airline a fee for landing at that airport, covering the cost of the airport airside infrastructure as discussed in Chapter 9 of this book. Here we saw that the landing fee is based on aircraft type specifics, like the aircraft's MTOW and noise and emissions characteristics, and on the time of landing. This landing fee is a *variable cost per flight*.

### 15.3.8 Navigation charge

ATM organizations charge the airline for the use of their airspace and the ATM services provided, as discussed in Chapter 2. This charge varies per country and is a variable cost, varying with the distance flown. The charge is calculated on the basis of three ingredients. First comes the aircraft's *weight factor*, the square of the aircraft's MTOW rounded up in tons. Next is the *distance factor*, which is the shortest distance between the point of entering a country's airspace and the point of leaving it, divided by 2. The third factor is the *unit price* that can differ per country. By multiplying the weight factor, the distance factor and the unit factor we get the navigation charge, in many countries a sizable *variable operating cost*. When flying above the open sea in airspace not belonging to a country, the airline pays a fixed service charge to the coastal states that monitor the operations over the open sea.

Distance factor **X** Weight factor **X** Unit factor

*Distance factor = the linear distance between point of entering a fir and leaving a fir in* $\frac{\text{KM or NM}}{100}$

*Weight factor =* $\sqrt{\frac{\text{Certified MTOW roundup in tonnes}}{50}}$

*Unit factor = Defined by the State of the FIR*

**Figure 15.4** Navigation charge calculation

In some countries, the airline also pays a Terminal Navigation charge upon departing from an airport. Whereas the ATM cost of approaching an airport in a STAR is covered in the landing charge, the ATM guidance for using the SID can be charged separately, with a charge depending on the aircraft's MTOW.

Navigation charges are always paid in hard currencies.

## 15.3.9 Catering cost

Airlines provide in-flight services, like offering meals and drinks. The cost for this is calculated on a per passenger base, so we treat it as a *variable cost per passenger* and therefore we can also regard it as a marginal cost. But this is for calculation purposes only; the real cost for catering is more complex. Here again, when an airline has a catering production facility at its home base, most costs are fixed in local currencies, and the variable cost is in buying the ingredients, which can be bought against either local currencies or hard currencies. Most airlines buy the catering for their return flight at the destination airport, and this goes against hard currencies and indeed per passenger.

## 15.3.10 Variable crew cost

After a long-haul flight the crew remains at the destination. This incurs hotel cost and per diem cost per crew member. On short haul operations, many FSC airlines operate a very early morning flight from the destination to the home airport. Here also, the airline faces these costs per crew member. Airline operating networks whereby the aircraft does not return to the airport where it departed with a certain crew, this crew may need rest at another place than where it departed, resulting in overnight cost. These costs are the *variable* crew cost and come *per crew member per day*. An entire crew or individual crew members may well need to be positioned to another airport to resume duties. These positioning costs are also variable crew costs.

It is important to realize that these variable crew costs are only a small fraction of the total crew cost. In Section 15.3.2 we saw that most crew costs are fixed costs that need to be divided by the flight hour production of these crews. As such we can understand that crews staying at the destination do not fly during that stay and thus dilute their annual productivity. The result is that the total fixed crew cost needs to be divided by a lower number of flight hours. Crews resting at outstations thus dilute productivity and increase the fixed crew cost on a per flight hour basis.

## 15.4 Indirect operating cost

Airlines are complex organizations and incur many costs that cannot be directly linked to the production. These are the indirect costs. At the end of a period, be it a quarter or a year, an airline needs to add up all these costs and divide them by the total ASK production by that period. The result is the IOC per ASK. The airline can calculate its IOC per flight by multiplying the IOC per ASK by the number of ASKs produced by that flight. The IOC per ASK is in itself an important airline KPI for managerial effectiveness. By becoming lean and mean as an organization, and boosting production, an airline derives a low IOC per ASK. As the indirect cost differ profoundly per airline, this IOC per ASK is an important competitive discriminator.

### 15.4.1 Overhead

The complexity and the geography that comes with aircraft operations require effective management involvement. The resulting layers of management generate indirect cost. Apart from the operation, an airline needs a robust Human Resource management, financial management and IT management. All generate costs that cannot be linked directly to the production and are therefore indirect costs, often referred to as overhead.

### 15.4.2 Cost of sales

Remember what was discussed in Section 11.3: selling tickets is the most important task of an airline. This generates significant indirect costs. Communicating with markets require advertisement campaigns. Selling tickets either go through the internet or through sales offices, and those all generate costs. All costs associated with selling tickets are sales costs or costs of sales.

### 15.4.3 Miscellaneous

Indirect costs also come in the form of insurance policies for liabilities, as airlines are by law liable not only to their passengers but also to damage on the ground resulting from a fatality. In some areas of the world, like the EU, airlines are also liable for inexcusable delays of their flights. These airlines need to cater for this DBC (Denied Boarding Compensation).

In established carriers pension liabilities to retired staff can form a significant indirect cost.

## 15.5 Airline revenues

Airlines generate revenues to cover all the cost that we discussed in the previous paragraphs, whereby the revenues should be higher than the total cost in order to be profitable. These revenues can come in various forms.

### 15.5.1 Fares

The main stream of income is in the fares that the market pays for transportation. Passengers, freight forwarders and mail companies pay for their transport requirements. The fare that they pay is ideally higher than the cost for that transportation. When demand is high, this is feasible, but aircraft also need to operate at moments of low demand. In such an environment, the fare should cover all costs, and if even that is a challenge, the fare should at least cover the contribution margin, as discussed in Section 12.1.7, meaning that the fares cover the fixed cost. Airlines use Revenue Management for setting the fares. We will discuss this is the next chapter.

Regarding the fares it is also crucial that the revenue income stream be *ahead of the cost outflow* – the income should be available before the cost of operation needs to be paid. This ensures a *positive cash flow*. In case of a negative cash flow, the airline needs to finance its costs before these are covered by income. One of the attractions of ticket sales through the internet is that the passenger pays for the transportation well ahead of the actual production of it. On the other hand, in case of code shares, as discussed in the previous chapter, when a passenger buys his or her ticket at another airline, it is up to the code-share partners to reconcile these joint income flows as quickly as possible in order ensure a positive cash flow for both.

Relevant also for airlines that operate international is that the fares should be paid in *currencies that can be used for covering the cost*. In Section 12.3 we saw that cost come in a variety of hard currencies and local currencies, and the airline need to ensure that the income stream ideally is composed of a comparable mix of currencies.

### 15.5.2 Ancillary sales

Passengers pay a fare for transportation, but airlines increasingly charge the passenger separately for the individual elements that are part of their product to the passenger. Checked luggage, seat reservation and extra legroom, meals and drinks, IFE amenities and Wi-Fi are all elements that can be *charged separately*. These are all ancillary sales.

Airline reservation websites are increasingly connected with reservation systems of car rental providers and hotel chains. This enables the passenger to not only buy the airline ticket, but to buy the other elements of travel as well. The airline receives a *commission* for this, and these are also part of the ancillary sales.

Especially at LCCs, which attract the passenger with a low base fare, the ancillary revenue stream forms a sizable percentage of the total revenues.

### 15.5.3 Services to other airlines

Many, and certainly the larger, airlines have in-house capabilities for ground handling, maintenance and catering production, based at their home base. Many airlines provide these services to visiting airlines at their home base against a charge. Remember that many of the costs at the home base are fixed costs, and by increasing output we lower the fixed costs per unit. *Maintenance services, ground handling services and catering sales* can form a substantial income, coming from activities the airline performs at the home base for the own airline.

### 15.5.4 Aircraft trade

When an airline buys an aircraft, it becomes the *owner* of that aircraft. When, after a certain period of operation the airline decides to *sell* that aircraft, its actual market value may well be higher than the depreciated book value of that aircraft. In that case, the airline generates a *book profit* on selling that aircraft, and these incidental incomes can form a significant form of income to the airline. It is *non-operating* income.

## 15.6 Results

At the end of a certain period, typically every quarter, the airline calculates all costs and revenues, and publishes its results. These quarterly results are the airline's operational results, and at the end of the year these results are added up and are put against all costs, leading to the airline's net profit or loss.

### 15.6.1 Operational result

At the end of every quarter an airline calculates its operational result. All operational costs, direct and indirect, fixed and variable, are added up and are set against all operational revenues. The operational result indicates how the airline operation performs economically.

In finance this is called the EBITAR (*Earnings Before Interest, Taxation, Amortization and Rent*). This result will vary per quarter as demand for aviation fluctuates per season. Operational results of the high seasons should compensate for lesser results in low seasons.

### 15.6.2 Finance

At the end of all four quarters the airline makes a total of the operational result *per annum*. This can constitute a calendar year but many airlines calculate on the basis of a *fiscal year*. Now the results of *financial activities* are added. Costs of loans and rents other than for the fleet are added. Airlines may have *a shareholder stake* in other enterprises and their gain is added. Airlines have revenues in various *currencies* and costs to be paid in mainly hard currencies. Currency losses or gains are added to the calculation. Airlines may also have sold aircraft for higher than their book value; this difference is added as well. All operational results and all non-operational finance costs or gains are added; the result is the airline's *gross* profit or loss.

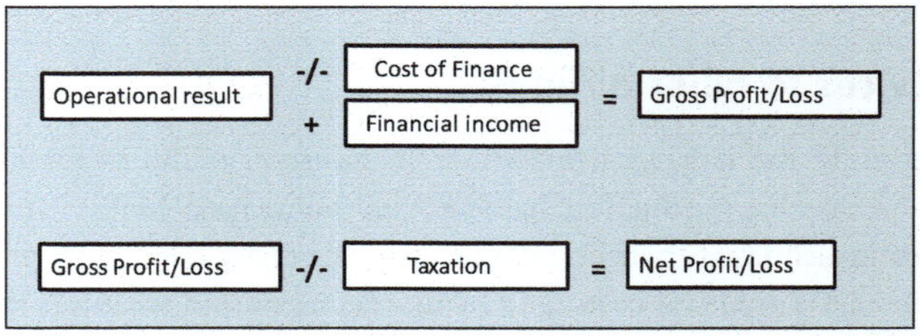


**Figure 15.5** Airline financial KPIs

### 15.6.3 Taxation

Gross profits are eligible for taxation, so the tax payable is subtracted from the gross profit/ loss. In Section 12.3.1 we discussed aircraft ownership and we saw that airlines purchasing their fleet *depreciate* this investment over time. We mentioned that depreciation is not a cash expense but an administrative activity, and here the relation with taxation emerges. Depreciation is not an expense but is regarded as a cost and therefore deductible for taxation. By depreciating fast, airlines can keep their gross profit low for taxation.

### 15.6.4 Net result

The final result after all the activities above is the airline's *net profit* or *loss*. Part of this profit can be paid as dividend to the shareholders, or can be added to the airline's equity on the balance sheet. By doing the latter, the airline increases its *solvability*, enabling the airline to finance investments in fleet, equipment or, as the very strong players like to do, stakes in other airlines. And most important for longer term survival in the cyclic industry, increasing solvability in good times is necessary to cope with bad times.

## Suggestions for further reading

Stephen Holloway: *Straight and level, practical airline economics*. Routledge, 2017.

CHAPTER 16

# Airline planning

In this chapter, we will discuss how airlines plan their operations on a strategic and tactical level. Strategic planning is where decisions are made that cannot be altered overnight. Decisions on what business model(s) the airline applies, what customers it serves, which connections it offers and with what fleet it operates. For practical reasons, we will discuss the elements of airline planning separately in the coming sections, but in real life these various elements are interrelated, and all with their own constraints, making airline planning a compromise by definition.

## 16.1 Market research

Airline planning starts with *defining attainable markets*. In Chapter 4 we saw that airline markets can be divided between business travellers, EFVR travellers and leisure travellers. Starting point for every market assessment are the demand potentials from, to or via the home airport(s) of the airline. In Chapter 6 we saw that airports have demand potentials on the basis of location and catchment area. The airline needs to translate these potentials to connections by defining destinations from the home airport. This is performed by market research.

Actual demand between airports is found in databases as provided by various data providers. These data reveal historic and actual data on the number of tickets sold for non-stop and transfer flights and by which airline. On the basis of these data an airline can assess *gross* demand for connections and market shares of the airlines that produce these connections. In market research, we call these connections *city-pairs*. An existing connection is an existing city-pair. Some databases also indicate an average fare level for which these tickets were sold. With these data, quantitative mathematical analyses can be performed, revealing market potentials for a certain route.

Setbacks of this approach is that the data are purely quantitative, telling little or nothing about connections that are not offered yet, nor the type of passenger or the price elasticity of demand. This implies that for starting new connections, a degree of *trial-and-error* cannot be avoided. And here we see airlines dropping destinations regularly shortly after the introduction. Despite thorough quantitative analysis, serving the destination may in real life not bring the expected number of passengers.

For analyzing existing connections, it implies that the mix between the three markets needs careful qualitative consideration. If a route primarily relies on leisure markets for filling the aircraft, the route may financially underperform due to a lack of business- or EVFR passengers, and become vulnerable to seasonal demand fluctuation.

The most important element, though difficult to quantify, is the *attainable market* on the basis of gross figures and competitor's market shares only. The airline needs to make a *qualitative judgement* on what percentage of the gross market it can take, on the basis of a fair *competition analysis*. What are the competitive forces of the other airlines operating the route, and how is the own airline judged by the markets against these competitors?

Hub airlines also consider the added value of a new connection within its hub network.

As many airlines also transport belly cargo on their long-haul routes, *cargo potentials* enter the equation as well for selecting destinations in the network.

## 16.2 Network planning

An airline bases its selection of destinations on the basis of market analysis. These destinations are connected with the home airport on the basis of O+D and/or hub potential of the home airport.

The *first* element that the airline needs to ensure is a *legal endorsement* for flying to that destination on the basis of the freedoms of the air, as we discussed in Chapter 2.

The *second* element to consider is the *distance* between the home airport and the destination. It can be short-, medium- or long-haul, and it defines *what kind of aircraft* we need to apply and what operational planning we need for this. In Chapter 11 we saw that long-haul aircraft come in high payload capacities and based on this capacity, we need to define the *frequency* level of the connection, and the need for feed at the home airport or/and the destination airport. Here we truly see the paramount importance of the *market potentials of the home airport* as the basis for an airline's network.

The *third* element to consider is the *type of market* for which the connection is valuable. Based on DESTEP analyses of the home airport and the destination airport, as was discussed in Chapter 7, a connection may be valuable to the business traveller, the EVFR traveller or the leisure traveller, with the connections between large airports containing a mix of the three. Based on this market mix, the airline defines its frequency level and choice of booking classes: fist, business and (premium) economy. This has implications both for the production cost of the operation and the income potential of the operation. Indeed, network planning is the core of the airline strategy and the airline's business model.

### 16.2.1 O+D networks

Airlines based at large home-airports can serve different markets with a mix of short- and long-haul connections, operated by a mixed fleet of narrow and wide bodies. The higher the demand from- and to the home base is, the more the airline can fill its wide-bodies at the home airport. With new long-haul aircraft types being more compact than their

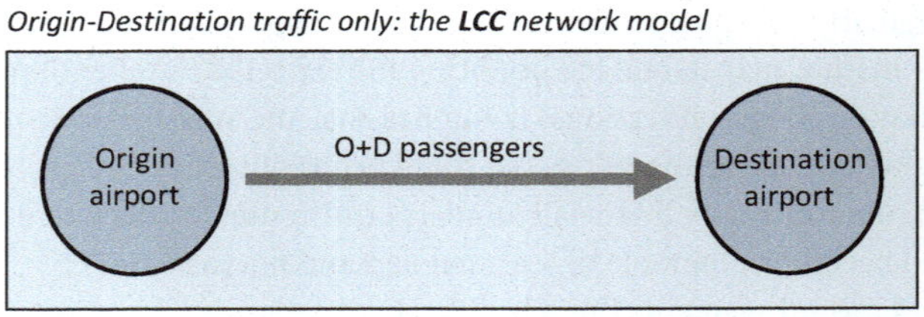


**Figure 16.1** Origin-Destination only: the LCC network model

predecessors, we see long-haul O+D networks growing. Non-stop connections are favoured by the regular premium passenger, ensuring a high yield, and can be produced by operating one flight only, ensuring the cost effectiveness of producing the connection. Combined with liberal and increasingly multilateral Transport Agreements, we see the development of long-haul O+D networks, like on the North-Atlantic- or the Pacific airline markets.

LCCs have developed O+D networks in their continent of operation. By exploring demand to and from regional airports often underserved by FSCs, LCCs have expanded the number of continental city-pairs served exponentially. The combination of liberal rules, economic growth, low production cost and low fares have developed aviation markets that did not exist before. Another factor boosting O+D networks is found in the constant process of urbanization in the world. Here we see formerly regional cities developing into metropoles, think of Manchester of Bangalore, justifying direct connections to other cities.

Regional O+D airlines may not operate from airport A to airport B and then return to airport A, as is common to many networks. Instead, they fly from A to B, and then, on the base of actual market demand at airport B, they fly from B to C, from C to D. An issue faced by this kind of network is that the airline may have resting crews scattered over the network, lowering crew efficiencies.

## 16.2.2 Hub networks

If O+D demand is insufficient for filling the long-haul wide-body aircraft, the airline needs a feeder network to the home airport by applying a *hub system*. Airlines based at secondary airports certainly need a hub feed if the airline operates wide-bodies to long-haul destinations. These airlines may need a feed at the destination airport as well in order to fill the large aircraft, the rationale for airline alliances, as we saw in Chapter 14. Through airline alliances, airlines connect their feed systems to the long-haul systems of the partners. And also, hub airlines fly primarily to hub destination airports. Airlines with large O+D demand can fly long-haul to secondary destinations, thereby offering more non-stop connections to the business markets.

Hub networks were first developed by the US domestic carriers for the reason that the US market has so many destinations to serve that could never be operated on an O+D basis only. The rationale for the hub system here is that by connecting flights at the home airport, a *multitude of one-stop city pairs* arise. Already early in the development of hub systems, these US carriers developed multi-hub networks, consisting of a network of hub home bases. We now see many large airlines also outside the US developing multi-hub systems, allowing such airlines to distribute its passengers over the network.

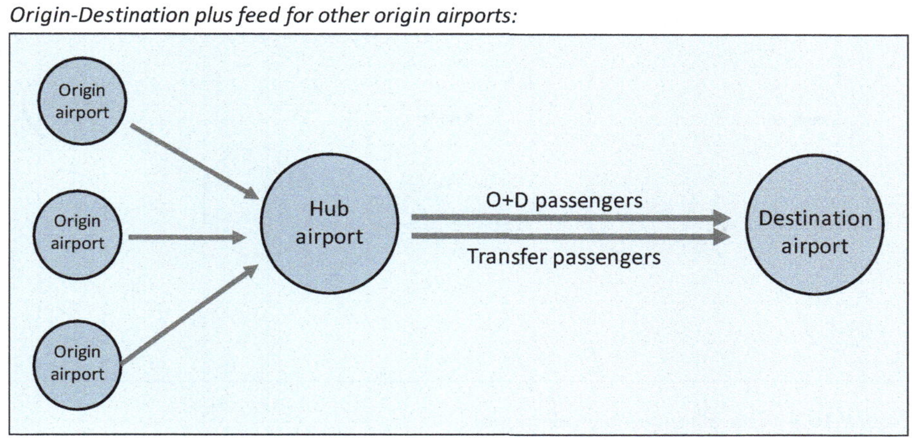


**Figure 16.2** The FSC network model

A hub connection is the most logical choice for city pair connections with demand that does *not justify a non-stop connection* at reasonable frequencies. Here we have the economic rationale for a hub connection. But we also see here the vulnerability of that hub connection: as soon as the market grows to an extent that the non-stop connection is feasible, the hub connection will lose its attractiveness.

A hub system has two hidden cost: *lower fleet utilization* and *higher production cost* of the connection. Lower fleet utilization results from aircraft scheduled to allow the passenger to transfer at the home airport. This leads to longer TATs for the aircraft and therefore to lower flight hour production. The higher production cost results from the fact that two flights are needed for the connection, as well as a relatively costly operation at the home airport to ensure the connection.

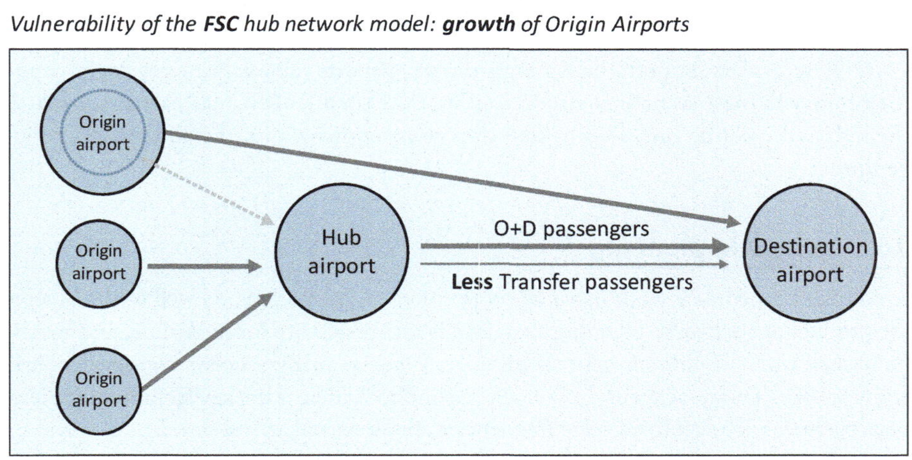


**Figure 16.3** Vulnerability of the hub model

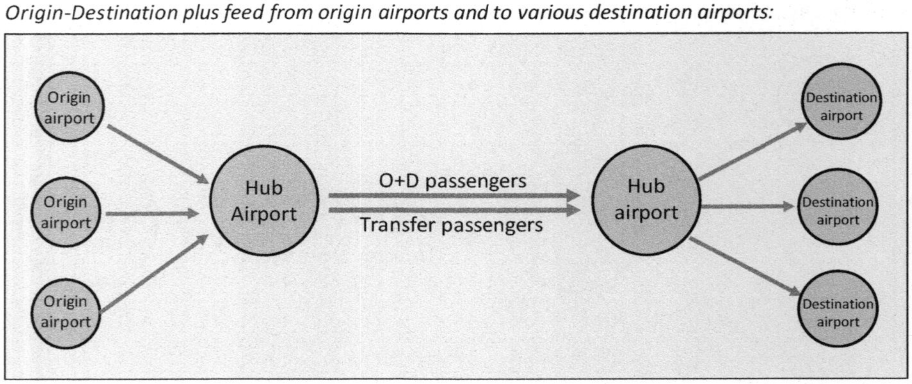


**Figure 16.4** The alliance network model

Feeder networks often have a third setback in the form of imbalance of demand, the demand difference between the outbound- and the inbound flight that we discussed in Chapter 4. A flight meant to feed the hub from a destination airport needs to be flown to that airport first and perhaps at low demand. Here we see feeder networks sometimes producing surplus seats. Many FSCs use these surplus seats, resulting from demand imbalance, to compete against LCCs by offering the seats against low fares.

Based at a secondary airport airlines can choose to limit the network to short- to medium-haul destinations only. If the home airport is centrally located within the continent, the airline can develop a regional hub system. When an airline operates at very high frequencies, connectivity at the hub airport emerges without compromising on aircraft utilization.

Hub airports are nevertheless indispensable for generating connectivity, the ability to fly from anywhere to anywhere. This is generated by connecting hub networks, and this is exactly what airline alliances do. By connecting the home-hub airports of the various alliance members, global networks are created by which a passenger can indeed fly between virtually any two airports, with two, or often only one, transfer point.

With secondary airports becoming primary airports, alliances increasingly connect their hub with these secondary airports within the network of alliance partners, bypassing the hub and creating non-stop or one-stop connections within the alliance or the joint venture.

## 16.3 Schedule planning

An airline's schedule is both its value proposition to the markets, as well as the blueprint for production. Schedule planning therefore is a core activity for an airline, and needs to be *aligned* with the airline's network strategy. Whereas many scholars consider an airline connection as a homogeneous product, the airline's schedule is the key factor that makes it a heterogeneous product by offering frequencies and preferred arrival times.

Connections aimed at business travellers need frequencies, whereby the airline with the highest frequency often has a disproportionally higher market share than the airline with lower frequencies. Traditionally high frequencies were a key factor for short-haul

*Development of **the Alliance and Joint Venture** network model*

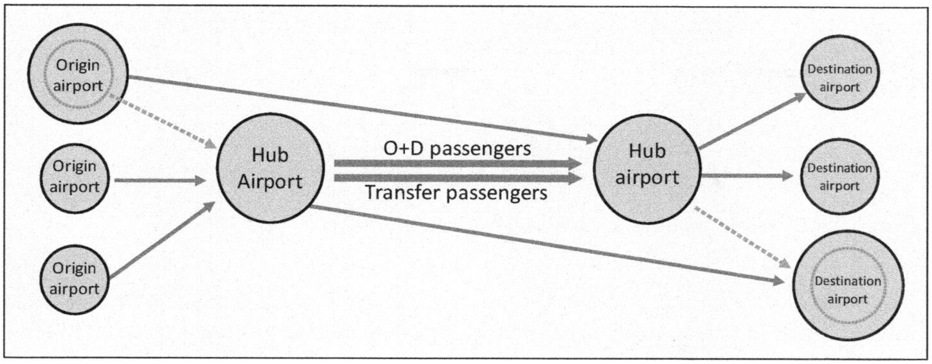


**Figure 16.5** Development of the alliance and JV network model

connections only, but in today's airline industry frequencies are paramount for attracting the business traveller on long-haul connections as well.

Scheduling needs to follow the demand variations, as discussed in Chapter 4, like offering same-day returns for business travellers on short-haul trips.

Scheduling is also the core of hub systems. An airline connecting its flights at the home base is doing so by scheduling its flights in transfer banks. Airline schedulers plan their schedule in *Universal Time Co-ordinated* (UTC) – the standard time based on an atomic clock, basically the same as Greenwich Mean Time – and translate this to local times at the departure and arrival airport.

In scheduling an airline we consider the flight to the destination and the return flight, combined with the turnaround time at the destination airport, and we call this a *rotation*. Here the importance of flight times enters the equation, as these flight times and the required turnaround time at the destination airport define the time length of a rotation. For short-haul operations, the TAT at a busy destination airport can have great impact on the rotation time. This time length indicates after how much time the aircraft returns to the home base, available for a next flight.

When designing the schedule the first thing that the airline needs to ensure is the availability of an *airport slot* for operating the flight at the desired frequency and schedule time. In Chapter 8 we saw that airports increasingly get congested in number of ATMs, and slot availability becomes an increasing constraint for operating an optimal schedule.

With the range capabilities of long-haul aircraft today, it has become common to fly non-stop to the destination and return to the home base from there, *the single-destination* operation. Combining more destination in one rotation, common in the old days, has lost importance. The emergence of ultra-long-haul flights leads to the complication that actual flying time may vary significantly in a wide range, leading to flights arriving too early or too late at the destination airport.

### 16.3.1 Block times

We just mentioned the TAT time at the airport. This includes the required taxi-out time at the departure airport, as well as the required taxi-in time at the destination airport. When

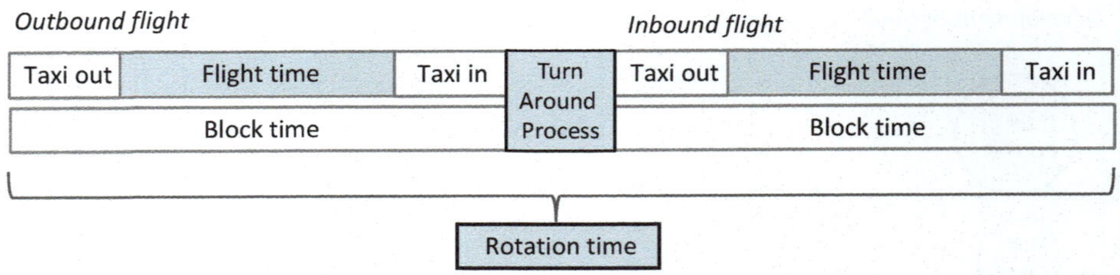


**Figure 16.6** Flight time, block time and rotation time

the required *taxi times* are added to the flight time, the result is referred to as the block-time for that flight. In scheduling, we use these block times and not just flight times. Two flights to different destination with identical flight times may have different block times, due to the size differences between the two airports. Two flights to the same destination may have different block times due to the differences in congestion during the day of one or both of the airports.

As the taxiing, as part of the block time, is done with the engines running, engine maintenance planners also use these block hours instead of flight hours in order to plan engine maintenance.

### 16.3.2 Seasons

Many airlines operate two schedules per year, the summer schedule versus the winter schedule. Airlines do this for three reasons: the shift from winter time to daylight saving time, the seasonal fluctuations in demand and differences in wind pattern resulting in different flying times between winter and summer.

Every year in spring, most countries in the world apply daylight saving time by setting the clock one hour in advance. For airline schedules this has the impact of all schedules times shifting one hour up from UTC. From here, airlines operate their summer schedule. In October of that year, these countries turn back the clock with one hour, whereby airlines start operating their winter schedule. These two schedules are referred to as the IATA schedules.

Airlines may see demand differences between winter and summer and may adapt their schedules to this demand variation in terms of aircraft type deployed, frequencies or operating a destination on a seasonal basis only.

## 16.4 Strategic fleet planning

In strategic fleet planning we need to consider two elements: fleet composition – what aircraft types the airline has in its fleet – and fleet size – how many of a type the airline operates.

The choice of aircraft types is defined by the distance to the destination, and what range capability is required to reach it. In Chapter 10 we saw that the range capability of an aircraft type defines its payload capability as well, so for long-haul flights we need wide-body aircraft types, and for short-haul operations we look at narrow-body types.

Strategic fleet planning is *long-term planning*; the costs of introducing aircraft into the fleet are high and need to be earned back over time. Nevertheless, this long-term planning needs to be *flexible* for coping with future market developments, and it needs to be *robust* in terms of aircraft life-cycle, avoiding the risk of economic obsolesce, as discussed in Chapter 11.

### 16.4.1 Aircraft size

Now we see a direct link with market research and scheduling: based on the anticipated demand for a city-pair and the required frequency level, an airline needs to ensure that the payload capacity of the proposed aircraft *matches the anticipated demand* on a per rotation basis, for passenger payload and, if required, for cargo payload as well. The airline needs to ensure that the capacity matches the *planning load factor*, the load factor that ensures a profitable operation. If the aircraft is too large for this demand per flight, the airline may need to limit its frequency level, albeit against the detriment of its desired market position in business markets. This implies that certainly long-haul city-pairs need to be attractive for other markets as well in order to ensure a profitable load factor on a per flight basis. Here we see the rationale for more compact long-haul aircraft types, enabling airlines to operate long-haul city-pairs in high frequencies with a moderate need for passenger quantity per flight.

When the actual load factor becomes larger than the planning load factor, the aircraft becomes too small for the demand. Airlines want to avoid this because now they cannot sell a ticket to a passenger who wanted to fly, forcing that passenger to either book a flight on another flight or booking a flight with a competing airline. This is called *spill*, a figure that is sometimes difficult to quantify exactly.

The main challenge in capacity matching demand per flight is in the wide variations in demand, as we discussed in Chapter 4. Demand fluctuations, imbalance and volatility make that airlines seek optimal flexibility in capacity deployed on a per flight basis. This is the rationale for aircraft family types, as discussed in Chapter 10. If the capacity increments within the family type are considered as insufficient, the airline may decide to deploy various aircraft types to a destination as per the anticipated demand variation during the day, the week or the season.

### 16.4.2 Fleet size

The number of aircraft, the *fleet size*, that the airline needs for operating the network in accordance with the schedule is defined by the total of destinations served in the network, the rotation times for these destinations and the frequency levels by which the destinations are served. This total defines the *required fleet size*. Airlines serving many destinations at high frequencies require large fleets. Airlines serving destinations at different distances require a fleet mix of various aircraft types. Airlines operating a network with comparable distances may be able to operate a single-type fleet. Airlines concentrating on O+D only, thereby not offering hub connectivity, can plan their schedule denser, lowering the required fleet size. Here we see the importance of aircraft utilization as airline KPI, as discussed in Chapter 15: the higher the airline can utilize its aircraft, the more the airline can produce with a fleet. This is key for cost efficiency as every aircraft in the fleet comes with huge fixed cost.

In defining the required fleet size, an airline needs to define a minimum fleet size. In Chapter 12 we saw that aircraft generate type-related cost on crew training and maintenance operations, and these costs need to be spread over a certain minimum number of aircraft of that type. It is hard to define this minimum as it depends on the cost structure of the airline, but as a rule of thumb a fleet of ten aircraft is generally considered as a minimum.

Airlines with large fleets can vary actual capacity within the fleet of the same type. But even within the same aircraft type version, airlines may vary actual seating capacity per service class for optimizing the fleet to actual demand per market. The type assigned to business destinations may have a larger business class installed, and aircraft of the same type operating to leisure destinations may have a denser seating configuration.

### 16.4.3 Fleet expansion

A special element of fleet planning is in fleet expansion. When business goes well and demand is growing, the need for additional capacity will rise. The problem here is that demand will grow with a certain percentage and perhaps unequally divided over the network. When another aircraft of a certain capacity is added to the fleet, the fleet capacity goes up with the capacity of the new aircraft, but it may not match the actual need for expansion. If an airline with ten aircraft sees a 5% growth on certain destinations and decides to add an 11th aircraft to the fleet, its capacity goes up with 10% in one single aircraft, not matching the 5% on only some routes. This "growing pain" in fleet expansion is a prime cause for constant overcapacity at many airlines.

## 16.5 Crew planning

With a flight schedule in place, and an aircraft type assigned to the flight, the airline can plan the crew for a flight. A crew consists of cockpit crew, the pilots, and cabin crew, the flight attendants, and crew composition is subject to *tight rules and regulations*, both in quantity (the required number of crew members) and in quality (licensing and flight hour experience), as we saw in Chapter 14. Above all, crews consist of humans and humans get tired after a certain period of working, the basis for tight rules and regulations on crew deployment, the duty-time regulations.

The more the airline utilizes its aircraft, the more crews it will need to operate the fleet. Crews generate sizable fixed costs to be spread over as many flight hours as possible. Here we have the ingredients for one of the most complicated aspects of airline planning, crew planning.

### 16.5.1 Crew ratio

The *crew ratio*, the number of crews per aircraft, is the KPI for crew planning quality. The required crew ratio depends on the type of operation. For short-haul operations, the crew remains with the aircraft at the destination airport and returns with the aircraft. This results in a low crew ratio, at many airlines between 1:5 and 1:7. The crew ratio goes up when short-haul aircraft are stationed at a destination airport overnight. Due to the duty-time regulations, the crew that flew the aircraft to that destination is not allowed to operate the early morning flight to the home airport, resulting in overnight stay of two crews.

When flying long-haul, the crew ratio goes up quickly to around 1:10. After arriving at the long-haul destination, the crew needs to rest. A fresh crew will operate the return flight and the arriving crew remains at the destination for at least 24 hours of rest. This period goes up as the crew has passed more time-zones during the flight. This crew can fly back home only after the required rest period.

### 16.5.2 Capacity flexibility

An airline may want to deploy different capacities to a destination in order to cope with demand variation, but this has consequences for the crew ratio. The rules and regulations stipulate that a cockpit crew can only operate the aircraft type for which the pilots have a type-rating. Family type aircraft come with a common type-rating for increased flexibility in crew planning. Cabin crew members are allowed to operate many different aircraft types, but here we have the planning limitation that a cabin crew needs at least to consist of a cabin attendant for every 50 seats in the aircraft. As the airline may want to deploy different capacities to a destination in order to cope with demand variation without compromising on crew efficiency, they have the advantage of a common type rating for cockpit crew against the disadvantage of a cabin crew that is larger than required on the flight with the smaller capacity. Many airlines therefore deploy preferably one but never more than two different aircraft types on long-haul flights that are operated daily or more. For flights with lower frequencies a single aircraft type is favoured because the advantage of capacity flexibility is outweighed by the need for crew efficiency.

### 16.5.3 Fleet expansion

With the expansion of a fleet comes the challenge of the need for new crew members. Both cockpit crew and cabin crew need, according to regulations, to have a variety of experience. This means that an airline can plan a low experience young first officer only next to a very experienced captain, or vice-versa, and a set of cabin attendants governed by a purser also with a minimum average of experience. Airlines need to develop that experience level in-house; they need staff loyal to their company to remain employed with the airline for a long period.

It is fair to say that crew planning is often more challenging than fleet planning, and flight cancellations occur due to unavailability of a crew, and not of the aircraft.

## 16.6 Maintenance planning

We stated that the required fleet size results from the number of destinations, the frequencies and the rotation times, but there is a final element that needs to be included in the equation as well, the *anticipated down time for maintenance* of the aircraft. We saw in Chapter 11 that aircraft maintenance is to a very large degree plannable, with check intervals and check durations well defined. This implies that proper airline management, based on proper fleet planning, also needs a very proper maintenance planning. *Equalized maintenance* can be applied to ensure maximum aircraft availability in periods of high market demand against maintenance activities planned at periods of low demand. When this maintenance is outsourced and executed at an airport other than the airline home airport, the ferry flights to and from the maintenance base need to be assessed as well.

As planned maintenance is defined in flight hours, and as airlines will plan their aircraft in such a manner that all accumulate a comparable number of flight hours, airlines need to properly plan the introduction of new aircraft. It is not wise to introduce many aircraft at the same time as they will all be due for maintenance at the same time as well. A proper fleet introduction ensures a certain time increment between aircraft in the fleet becoming due for maintenance.

Airlines also need to ensure the availability of line maintenance, both at the home airport but certainly at the destination airport. A technical problem at the destination airport can be a prime cause for schedule disruption, reason for airlines to ensure a maximum technical integrity of the aircraft at the departure from the home airport.

Here also, deploying various aircraft types to a destination has its repercussions. Line maintenance at the outstation needs the ability to cope with technical failures of the aircraft, and deploying different aircraft types can complicate this. Spare parts as well as licensed engineers need to be in place for all aircraft types deployed to that airport. Preferably airlines fly to destinations where a home carrier operates the same aircraft type, with full capabilities to assist when necessary. In real life, this ability to assist is sometimes limited by the fact that aircraft of the same type can differ profoundly in engines or components, often selectable, as we saw in Chapter 10.

## 16.7 Commercial planning

At the start of this chapter we discussed attainable market sizes for a specific city-pair. What we did not discuss there is the question how this market size is divided between demand from the home airport or demand from the destination airport. In leisure markets this demand is very one-sided: all demand comes from the home airport as there hardly is any demand at the leisure destination. On the other hand, on city-pairs with demand from business- and EVFR markets, this demand often is two sided: demand from the home airport to the destination can well equal the demand from the destination airport to the home airport. In all cases an airline needs to assess where demand for the connection comes from, because airlines often need to sell tickets at both sides for profitability. Here, commercial planning becomes vital: how and where does the airline sells its tickets, and against which fares?

In Chapter 14 we discussed the marketing challenge for airlines, market recognition and market acceptation, in combination leading to market appreciation. Selling tickets not only in the home country of the airline, but at all destinations as well, require market communications in the form of advertisements and presence at social media. Large airlines may well sponsor sport teams for generating market awareness. The aim is, in all cases, to seduce the passenger to go to the website of the airline for buying a ticket, or to be accepted by that passenger on a generic website. A passenger needs to be attracted to buy a ticket in the reservation system of the airline.

### 16.7.1 CRS

Computer Reservation Systems are vital for selling airline tickets. These systems provide data on the availability of seats on a per flight basis. When combined, where many airlines are connected to the same system, including hotels and car rental companies, we call them GDS (Global Distribution Systems). Originated by initiatives performed by individual airlines, the suppliers of GDS have become organizations independent from the airlines. Amadeus, Galileo and Sabre are examples of this.

Key for airlines is to be on top of the retrieve result when a passenger or a travel agency seeks a connection on a website. This can be the website of an airline, an alliance or an independent generic ticket website. They are all connected to the GDS systems.

### 16.7.2 Fare levels

Airlines have in most cases the freedom to set their fares, and may to a certain degree, synchronize their fares with airlines they cooperate with. Airlines should set their fares at levels that guarantee the cost of the airline to be covered by the revenue stream. If this is not possible, in cases of low demand, airlines should at least cover the contribution margin, as we discussed in Chapter 15. An airline that consistently sets its fares below its cost, will incur losses and ultimately bankruptcy. This is relevant as many define an LCC as an airline offering low fares. Many airlines that offer low fares do not operate on a low-cost basis at all.

Attainable fare levels differ per market segment. In the premium booking classes, high fares are attainable dependent on the perceived customer value of the airline brand. But here also the passenger may select the lowest fare for a comparable connection. EVFR- and certainly leisure markets are very sensitive to fares. In the lowest booking classes, certainly on short- and medium-haul flights, fare level is the first selection criterion of many passengers.

There is thus good reason for airlines to really plan their fare strategy, based on perceived quality, choices available to the various markets and the influence of demand fluctuations.

### 16.7.3 Revenue management

Most airlines do not fix their fares but start from a base fare. From here, the airline discriminates fares based on various restrictions, in order to get the right fare from the right market. The week return with a weekend in between is for the leisure traveller at a low fare, the day return to the same destination, aimed at the business traveller, goes against a higher fare.

Next to this, the airline sets the fare on an individual base to a passenger on the basis of the timing of the flight and the expected demand for that flight. This is done by revenue management, executed by systems connected to the GDS systems. On the basis of actual demand for an individual flight, the fare for that flight may go up or down. Logarithms on the basis of historical data predict last minute demand, avoiding the airline to sell tickets against a lower fare far ahead of the flight than attainable at a shorter period before the flight.

Revenue management enables the airline to influence actual demand and can be used to distribute passengers over a multi-hub network or to manage the income stream in terms of various currencies. Revenue management makes the fare setting a form of on-line auction for every individual ticket sold against the highest attainable fare.

### 16.7.4 Ancillary sales

One of the innovations that LCCs introduced is the notion of the passenger paying only for what they want. Whereas legacy carriers offered all product elements in one fare, the LCC distinguished between the base fare, the fare for the travel only, and service elements like checking in luggage, seat reservation, or meals and drinks on board. These elements of the product are priced separately, and the resulting income stream is called ancillary sales.

This is based on the observation that in the lower booking classes, consumers base their decision on price. By offering the lowest fare the passenger is attracted. Only after the booking is made, the passenger chooses to add service elements to the reservation. The end result may be a total fare comparable to other airlines. As this is basic consumer behaviour, we see FSCs increasingly doing the same thing, the reason being that ancillary income streams are generated by the FSCs as well.

## 16.8 Modelling and simulating

When looking at all the strategic elements that we discussed in this chapter, it has become obvious that airlines cannot make decisions on the basis of trial-and-error or, as in the past, on the basis of gut feeling or implicit assumptions. Strategic decisions need to be based on aligning market research, network planning, fleet planning, scheduling and commercial planning, whereby the end result is always a compromise. Finding the optimum here needs modelling and simulation, whereby the effects and consequences of all aspects of airline strategic planning become visible in the what-if scenarios. Through modelling and simulation, airlines can find out what the sensitivities in their strategy are, where they could be vulnerable but also, where the opportunities for profit and growth are.

## Suggestions for further reading

Ahmed Abdelghany, Khaled Abdelghany: *Modelling applications in the airline industry*. Ashgate, 2009.

Massoud Bazargan: *Airline operations and scheduling*. Routledge, 2010.

Philipp Goedeking: *Networks in aviation*. Springer, 2010.

Rigas Doganis: *Flying off course*. Routledge, 2014.

## Interesting websites

Anna.aero

Routesonline.com

# CHAPTER 17
# Airline operations

In the previous chapter, we discussed airline planning on a strategic level, where decisions are taken that have a long-term impact. In this chapter, we will concentrate on an airline's day-to-day operation, how airlines cope with alterations that were not envisaged at the strategic level. There are many events that an airline needs to anticipate, and there are many causes for a disruption of the planned operation. In this chapter, we will also look at operational activities performed by airlines that may not be considered as strategic in itself, but nevertheless are indispensable for airline operations, like ground handling operations or catering production. In short, in this chapter we will consider airline management at a tactical and operational level.

## 17.1 Tactical fleet planning

In Chapter 4 we saw that demand for air transport fluctuates in many ways. In Chapter 11 we discussed the family-type of aircraft. Now, in tactical fleet planning, we see these elements coming together: based on anticipated demand fluctuations the family fleet is assigned to the various rotations in the network. The larger capacities are assigned to routes with high demand and vice versa. Demand fluctuates per destination on the basis of seasonality, and demand varies during the day based on actual demand at that time of the day. As such, airlines deploy a large capacity to relevant business destinations in the early morning and the early evening, with perhaps a rotation to a leisure destination in between.

In tactical fleet planning we see wide differences between airlines, based on fleet composition and based on planning flexibility. On fleet composition, we see airlines operating a single aircraft type where planning is straightforward. But most of these airlines operate a type in various capacities, and here decisions are made on deployment of the larger capacities versus the smaller variant. Airlines operating various aircraft types at high frequencies need to decide for which rotation during the day or week the various types are planned over the network.

At these airlines, we see differences in planning flexibility. Some airlines plan the fleet over the network at the beginning of the summer/winter schedule. Based on historic data and predictable seasonal demand variation, certainly the long-haul fleet is at most airlines planned in this manner.

Some airlines plan the aircraft a few weeks before departure, while others are capable of assigning a short- or medium-haul fleet on a per rotation basis only days before departure. These airlines use the booking data from the revenue management system to identify actual demand per leg of the rotation and adapt the deployment of the fleet to that. This "last-minute" adaption is made possible by the bank-scheduling of the fleet at the home airport resulting in departure- and arrival times around the same time, and by the fact that in short- or medium-haul operations the crew remain and return with the aircraft to the home base.

## 17.2 Crew pairing

Tactical fleet planning is closely linked to crew planning, as an aircraft can only operate when a crew for that aircraft is available. In tactical fleet planning, very often not the aircraft availability, but the crew availability is the limiting factor.

In the previous chapter, we learned about the crew ratio – the number of crews available per aircraft – a ratio to be kept as low as possible for cost reasons but sufficiently high to ensure an optimal deployment of the aircraft. It is a true challenge.

Airlines employ many staff as flight crew. All these individual employees need to be fit and certified for their duties. They need recurrent training and medical checks, and while flying they build up flight hour experience. They all need passports and valid visas, enjoy holidays and may have preferences for flying with certain colleagues or destinations.

All this vital information is stored in a database system on a per employee basis, and for every individual flight a crew pairing system composes a valid and certified crew for that flight.

The main discriminator in crew pairing is whether the crew operates the total rotation, and returns with the aircraft to the home base. This makes crew pairing relatively simple, ensuring high crew efficiency. LCCs operate this way; by deploying a multi-home network, crews return to their home base at the end of their duty time. This is very different at long-haul operations, where the crew remains at the destination and a "fresh" crew is required at the destination to fly the return leg of the rotation. With crews being aircraft-type related to a certain level, crew efficiency here requires a frequency level of the route enabling the redeployment of the crew shortly after the required rest period at the destination. On long-haul operations, fleet flexibility is limited due to this crew planning element and airlines need to make a careful trade-off between capacity flexibility versus crew efficiency.

## 17.3 Ground operations

So far, we discussed the planning of flights, but every flight starts on the ground at the airport. Airlines need to have their ground operations in place in order to actually execute all the flights that they plan. These operations involve passenger processing in the terminal, and aircraft handling at the ramp, with luggage handling in between. It needs to be performed at high quality and accuracy. Poor quality of ground handling leads to flight departure delays, and low accuracy leads to excessive cost for damaged or lost luggage and erosion of customer satisfaction. Ground operations are therefore a vital element of airline operations and are directly linked with other players in aviation, like airports, ground handlers and authorities at the airport. Much of what needs to be known on ground handling has been discussed in Chapter 10 on airport capacity management. Indeed, ground handling is an airline responsibility, but needs to be aligned with the capacity characteristics of the airport.

### 17.3.1 Turnaround process

Ground operations are responsible for the turnaround process of the aircraft. Upon arrival at the parking stand this TAT starts after the aircraft is on blocks and the engines are turned off. The aircraft gets connected to a ground power unit, the passenger doors may be opened and passenger disembarking can start. At the ramp a high loader or conveyor belt is positioned at the belly doors, and luggage located in the bellies behind the wing is unloaded. Only after that can cargo situated in the forward bellies be unloaded. This sequence ensures a quick availability of the luggage to be handled, already sorted for transfer or end-destination. It also ensures the aircraft not to become unstable, as would be the case if the forward bellies were unloaded first.

When the passengers have left the aircraft, catering can replenish the galleys and cleaning services can start cleaning the cabin. At the ramp, the aircraft can be fuelled for its next flight, water and waste services can do its job and eventual line-maintenance actions can be performed. Cargo for the next flight is loaded in the forward bellies and luggage of the departing passengers is loaded in the aft bellies. Meanwhile, after both catering and cleaning is done, boarding the passengers for the next flight is executed. After the doors are closed, the aircraft will be detached from its GPU, the pushback procedure is performed, and the engines are started up. This ends the TAT process.

Many airlines have simplified the TAT process by eliminating aspects like catering and sometimes fuelling in order to minimize the TAT. This is part of the LCC business model, where high aircraft utilization is the basis for the low cost-base. LCCs can turn around their narrow-body aircraft in 15–20 minutes. FSCs normally work with somewhat longer TTs on the narrow-body operations as arrival time adaption to the transfer bank defines the required departure time of the aircraft, not the shortest possible TAT.

On wide-body aircraft for long-haul operations, a minimum of some 90 minutes is required for the TAT process and here also, actual TAT times are often longer as the scheduled departure time is not set by the TAT time but by the required arrival time at the destination airport in another time zone.

### 17.3.2 Home-base operations

Many airlines operate their own ground operations at the home airport. In Chapter 7 we saw that many large airlines even own in full or partly their terminal at their home base(s).

Here, the airline has full control over the quality and accuracy of its ground operations, both for passenger- and luggage handling as well as for aircraft handling.

These operations require high investments in equipment, referred to as *Ground Service Equipment* (GSE). High loaders, conveyor belts, trucks with dollies, luggage carts, catering trucks, water & waste trucks, Ground Power Units, engine starters and pushback vehicles are part of the GSE that is needed for aircraft handling. Inside the terminal, often in the basement, luggage sorting and screening infrastructure is required. Inside the terminal, the airline may need check-in counters, customer services desks, self-boarding devices and luggage drop-off devices, all connected to the airline IT infrastructure. The airline may need a business class lounge, and hub airlines need transfer desks and services in the secured area of the terminal. Airlines need to buy, lease or rent all this for their home-base ground operations.

Both ramp handling and passenger handling inside the terminal also need many staff. Indeed, airline operations are labour intensive on the ground as well. The result is that ground handling at the home airport mainly consists of *fixed cost*, preferable spread over as many flights as possible.

Hub airlines operating in transfer banks have an additional challenge to cope with: they need their home-base ground operations to be able cope with *peak demand*, so the airline needs to plan its ground operations capacity for peak demand, resulting in obvious under-utilization in between these peak hours. The same applies to leisure carriers starting their operations in the early morning at the home base. Here we see a strong rationale for alliance airlines handling cooperation. More general, an airline may offer its ground handling operations to visiting carriers on a commercial basis, enhancing the cost efficiency of its ground operations and contributing to the revenue stream of the airline.

### 17.3.3 Outstation operations

Many airlines outsource their outstation TAT to a home carrier at that destination airport. This is based on contracts containing a service level agreement and goes against a cost per aircraft handled and a fee per passenger handled. At many airports, the home carrier offering its ground services to other airlines faces competition from independent *ground handling companies*. These ground handling companies are located at many airports and can offer their services to an airline at many or even all of its destination airports, whereas the home carrier can only offer services at its home airport(s).

The limitation of many ground handling companies is in the exclusion of line maintenance services. A ground handler needs to have a strong tie with an MRO outfit for offering line maintenance. In Chapter 13 we saw that MRO organizations need strict licensing, and at outstations airlines need repair parts coverage, elements that a ground handler cannot offer. So, for technical assistance, the visiting airline relies on the technical outfit of the home carrier. If the home carrier does not have line-maintenance capability of the visiting aircraft type, the visiting carrier needs to arrange this. Here we see independent MRO providers offering line maintenance services at airports, or visiting airlines collaborating on a joint line maintenance operation.

In all cases the *service level agreement* (SLA) is paramount; the airline needs to be ensured that the ground handling processes for its flight are in accordance with the quality and accuracy of its operations at the home airport. For the passenger, ground handling is a key element in product quality, and for the on-time performance TAT quality is key.

When an airline operates a high-frequency schedule to a destination airport, with many TATs to be performed at that airport, the airline may decide to operate the ground handling itself, perhaps in cooperation with (alliance) partner airlines. At major airports, we today even see joint ground operations performed by alliance airlines at airports where no partner is available.

## 17.4 Disruption management

No matter how robust a schedule is, how high-quality all planning and operational functions in an airline are, all airlines are faced with disruptions as there are many causes for disruptions, and most of them are outside the control of the airline. And as aviation operations require many players and many processes to perform in close concert, a single cause for disruption easily propagates throughout the system.

Disruptors can be categorized in various ways. They can be caused by congestion or by failure of infrastructure. They can be caused in the air, by technical malfunction or a passenger related event, or on the ground by weather conditions or events at the airport.

The likeliness of these disruptors is difficult to rank, as this depends on the season. In summertime, weather conditions may be generally better, but here congestion at the airports and in the air can become the highest in rank. In wintertime, adverse weather conditions may be the number-one disruptor. In all seasons, technical failures with aircraft or IT systems and passenger events in-flight, all of which can lead to an unscheduled landing en route, tend to cause disruption.

An important difference in disruptors is between a disruption that hits all airlines equally – like adverse weather conditions – and disruptions caused by an airline-related element hitting one airline only. In all cases, disruptions come with high costs: costs for booking passengers to other airlines, placing stranded passengers in hotels and leasing additional aircraft are examples. In the EU airlines are faced with DBC (Denied Boarding Compensation), stipulating that the passenger is entitled to financial compensation resulting from "avoidable" delays.

It can be concluded that it is impossible to avoid disruptions, and therefore airlines merely concentrate on managing disruptions with the aim to stabilize the operation and to bring back the operation to the schedule as quickly as possible.

### 17.4.1 Schedule robustness

The severity of a disruptor and the level of propagation of a disruption are first and foremost defined by the robustness of the schedule. When flights are scheduled tight for optimizing aircraft and crew utilization, a flight delay will easily propagate over the entire schedule for that aircraft. Therefore, airlines build "slack" in their schedules, like a longer TAT at the home base in the middle of the day, or longer scheduled block times for rotations later in the day and in the evening. These elements allow the airline to *absorb* a flight delay during the day with the aircraft returning in its schedule after the "slack" part. We call this *static* robustness.

*Dynamic* robustness requires more sophistication in schedule management. This is the ability to adapt all flights influenced by a disruptor in the system to a new schedule. Here, the operation is disrupted but as the new schedule times are known and communicated with the passengers, the damage of the disruption can be relatively low. This dynamic robustness is particularly important for coping with disruptions that will last for a while.

Key for this dynamic robustness is that crews are assigned to an aircraft. When one crew performs two rotations with the same aircraft, either a disruptor at the airports, with the aircraft or with one of its flights, will not quickly propagate to other rotations in the schedule. Here again we see the paramount importance of proper crew planning.

Long-haul operations are more vulnerable to disruption propagation than short haul. For this reason airlines have to prioritize long-haul on-time performance in case of airport disruptions' limiting the allowable number of ATMs. The longer the rotation time is, the more the delay will propagate into the planning for the aircraft and crew involved.

Dynamic robustness is particularly important for hub operations where a delayed flight results in passengers missing their connections. As soon as new schedule times are known and loaded in the system, passengers can be connected to their final destinations on the basis of these new schedule times.

### 17.4.2 Aircraft availability

Aircraft are vulnerable to technical malfunctions or failure of one of the many systems installed in that aircraft. An aircraft may start its rotation in excellent shape and arrives at the destination airport with a technical complaint. Any failed item not in the MEL is cause for an AOG. Items in the MEL may allow dispatch under MEL-relief conditions. If not, the aircraft is AOG (Aircraft On Ground) and needs to be repaired for the return flight. Here the robustness of line maintenance coverage enters the equation. Robustness in *quality* is required for fast and accurate trouble-shooting, finding the root-cause for the failure; robustness in *quantity* is required to ensure that an alternative part or component is available to restore the aircraft to its flying condition.

Crucial element here is the number of *different aircraft types* that an airline deploys to a destination airport. Every individual aircraft type needs coverage both in expertise and parts availability. Next to this is the necessity of the presence of a line maintenance provider with access to the required spare part. It is fair to state that an aircraft AOG at an outstation without maintenance coverage for that aircraft is about the worst-case scenario for an airline. Even when a flight is forced to make an emergency landing en route, the pilots will try to land that aircraft on an airport with facilities to handle the passengers and foremost to handle the technical problem of the aircraft in order to limit the propagation of the disruption.

When an aircraft scheduled for a rotation starting from the home base cannot be repaired within a short time frame, the airline may have a *spare* aircraft available at the home base for operating that rotation. This should preferably be an aircraft of the same type, allowing the scheduled crew to operate the flight. Otherwise another crew needs to be made available, reason for airlines to plan crew members on stand-by duty at the home base. When the aircraft is stuck at a destination airport, the reserve aircraft needs to be ferried to that airport with another crew. Again here, the availability of a spare aircraft may not be the main problem, but the availability of a crew may well be.

### 17.4.3 Operations control centre

Immediately when a disruption occurs, the airline needs to take action to cope with the disruption. This is done in the airline's Operational Control Centre, where ideally all operational airline functions are present. Here, integrated decisions on the disrupted operations can be made. The number of affected passengers and their transfer consequences, aircraft availability, crew availability and the planned schedule for aircraft and crew all add to the equation by which the OCC needs to ensure a dynamic robustness of the schedule, and the return to the original planned schedule as quickly as possible. The OCC is indeed the heart of the day-to-day airline operations.

## 17.5 Catering operations

To most students catering operations may not be regarded as appealing, as we merely concentrate on aircraft and airports instead of bothering about food and drinks.

Nevertheless, catering is very relevant for airlines since it is regarded as relevant by the passenger; it is very complex in its operations and involves significant cost. There is thus reason to spend a section on catering operations.

Certainly, in the higher booking classes and on long-haul flights, in-flight services, of which catering is part, is where passengers judge the quality of the airline product. Here, the airline needs to offer a menu of various meals, all freshly prepared and of high quality and coming with beverages of the same standard. Since airlines want to attract the regular passenger, these menus need to change regularly as well.

The larger airlines often have their own catering production facility at its home base. This is where catering for outbound flights is produced. A true logistical challenge for long-haul operations is to ensure the production of catering of the same quality for the return flight. This catering needs to be produced by a catering provider at the destination airport. This can be produced by the catering facility of the airline having its home at that airport. It can also be provided by an independent catering provider. They have the advantage of being present at many destination airports, enabling an airline to contract such independent provider at many of its destination airports.

## 17.6 Fuel preservation

All airlines operate the same aircraft types. Nevertheless, differences in fuel burn between airlines can exist, as there are various ways of preserving the actual fuel consumption in the operation of an airline.

This starts with *optimizing the OEW* of the aircraft on a per flight basis. On short-haul operations an airline can fill the potable water tank of the aircraft with some 800 kgs of water in the morning or, as airlines increasingly do, have it filled with an amount of water sufficient for that flight and replenish this at the TAT. This is possible since we saw that water services are not part of the critical path of the TAT. The same goes for catering; instead of filling the galleys on the basis of a full passenger payload, filling them on the basis of the *actual load factor* of the flight saves weight.

In flight operations, fuel can be preserved by *taxiing on one engine*. A *restricted use of thrust reverses* also preserves fuel. As airlines mostly operate to airports with long runways, there is no need for fast deceleration after touchdown. Ensuring an *optimal Centre of Gravity* of the aircraft, resulting from a proper loading of the aircraft's bellies, enhances the fuel efficiency for a flight.

Without doubt the most effective way of fuel preservation is in *on-time departure*. Flight plans are based on optimal speeds at optimal flight altitudes. A flight that departs too late has to make up for the time lost in-flight to ensure an on-time arrival, and departing too late may well lead to being forced by ATC to maintain a lower flying altitude for most of the flight, adding to the fuel consumed for that flight.

It can be concluded that most fuel preserving actions require discipline in operational procedures and quality of a timely TAT.

## 17.7 Cargo operations

So far, we discussed airline operations form the view of passenger operations. About 90% of all airline operations involve passenger operations. Many airlines, certainly on long-haul flights operated by wide-body aircraft, also carry cargo on passenger operations. Nearly 50% of all air cargo transported is carried as belly cargo on passenger operations. The remaining 50% is transported by dedicated freighter aircraft. In terms of cargo revenues, we see that roughly 90% is generated by dedicated freighter aircraft, implying that belly cargo transported by passenger operations only generate some 10% of the cargo revenue stream. Passenger airlines need to consider whether the payload requirement for cargo operations is justified by the revenues of cargo compared to the revenues of passenger operations.

Most of what has been discussed on airline operations in this chapter is valid for cargo operations as well. However, there are some significant differences between passenger operations and full cargo operations.

The first important difference is that *cargo is transported one-way* by definition. Passengers almost by definition return to the airport where they started their journey, buying return tickets, whereas cargo does not. This has far reaching implications for cargo networks. A cargo airline may have a stable demand for its services from A to B, but if there is no stable cargo demand from B to A, the cargo airline cannot operate a rotation A-B-A. Often, the cargo aircraft needs, after unloading its cargo at B, to make a positioning flight from B to C to collect the cargo payload for the flight back to A. Cargo aircraft therefore often operate in triangles. Flying empty, so without payload, is very costly; therefore the distance between B and C should be as short as possible. When very short, the forwarder can, in collaboration with the airline, decide to transport the cargo by road from C to B.

The second difference is in the airports served. The preferred destination airport for cargo may well be another airport, another location than the passenger airport. Certainly, with leisure airports we see a low demand for cargo operations. Cargo destinations concentrate on industrial areas or logistic cargo hubs, connected to cargo ground transport modes.

A striking difference is in the absence of passengers in cargo operations. This implies that the cargo airline does not need cabin crew, no in-flight services like catering and no clean cabin with seats, lavatories and in-flight entertainment installed. But, on the other hand, perishable cargo or life stock can have its own specific need for care during the flight. Without passengers, the dedicated freighter is subject to other rules and regulations, often more lenient towards to transport of hazardous or dangerous cargo.

As cargo does not complain about delays or late arrivals, it is easy to assume that cargo operations therefore are less sensitive to schedules. However, cargo operations are part of logistic chains whereby on-time arrival of the cargo can be crucial to a just-in-time logistic chain.

Where passengers, certainly in the higher booking classes, seek for the shortest travel times, resulting in airline offering non-stop connections, this fuel consuming way of operations is not required for cargo. In Part III we saw that cargo operators load their aircraft up to its MZFW and fly with en route refuelling stops to the final destination.

## 17.8 Planning IT

When reading this chapter, it becomes obvious that operational planning of airlines involves many interrelated aspects, whereby single disruptors to the operations can have a great impact on the day-to-day operations of the airline. Airline planning itself is closely tight to the planning processes of the airports that the airline is operating between. It is no surprise that these planning processes can only be executed by using sophisticated and dedicated software programmes. This has resulted in a handful of providers of integrated software packages for airline operations and a multitude of providers offering specific functionalities, either stand-alone or as plug-in to more generic software packages.

Most of the integrated software providers have their origin in airlines that originally developed software for their own use. Most notably Lufthansa Systems nowadays is one of the leading providers of airline- and airport planning software. Another large provider is Sabre, having its origin in American Airlines.

These providers offer aircraft scheduling programmes that integrate with crew pairing programmes. These packages can be connected to gate planning programmes used by airports, and be integrated by maintenance planning programmes.

Crucial for optimizing revenues is Revenue Management software. Here we see, as with many software packages, that optimizing the operations is not a matter of just buying a software package, the user of the package needs to fill in its own parameters in order to have the software doing what it is supposed to do. So, airlines need a profound understanding of how all that software works and what it can do for the operator versus what the operator has to do itself in order to benefit from the software.

## Suggestions for further reading

Cheng-Lung Wu: *Airline operations and delay management.* Routledge, 2010.

Gerald N. Cook, Bruce G. Billig: *Airline operations management.* Routledge, 2015.

Peter S. Morrell, Thomas Klein: *Moving boxes by air: the economics of international air cargo.* Routledge, 2018.

# PART V
# Epilogue

# CHAPTER 18

# Into the future

This textbook is about the fundamentals of aviation operations, but with one obvious limitation, it is about the fundamentals of *today*. This implies that by the time the student will enter the aviation labour market, various elements of this book may well be outdated in the constantly changing aviation industry. This textbook has tried to present all *fundamentals* in a generic manner, hoping that knowledge as presented will not become outdated too soon. And perhaps more important, by understanding the fundamentals of aviation, and the dynamics that changes the aviation industry constantly, the student should be well equipped to identify and understand the changes in his or her professional life.

Predicting the future is a problematic activity and seldom successful. The future may become defined by presently unknown influencers. A longer-term prediction of how commercial aviation will develop in the coming decades is full with these potholes. How the world will cope with climate change, how actual demand will develop with unpredictable saturation effects, the commercial viability of non-fossil fuel and development of surface transport modes are a few of these unknown factors that will have a far-reaching effect on the development of aviation. And as aviation strongly reacts on the world economy and global politics, the future of aviation is unknown as the future of the world.

On a shorter term, say the next five to ten years, predictions are made possible by understanding and carefully extrapolating present developments. And even doing this is inherently dangerous as every extrapolation is highly speculative by nature.

Taking all of the above in mind, we will in this last part of this textbook look into the coming years to see which fundamentals of aviation operations may change in the foreseeable future.

## 18.1 Demand growth

In Chapter 4 of this book we saw that the explosive growth of demand for aviation is driven by three factors: economic growth, the globalisation of societies and the constantly lowering air fares. The first factor, economic growth, leading to the growth of middle classes around the world, will continue, as will the globalization of society. The key factor for aviation is in the third growth factor, the low air fares. We see that certainly in the leisure markets, demand has a high price elasticity, with low fares leading to increasing demand.

The fundamental question for the future is to where this elasticity will go. At a certain low fare level, the other costs of travelling, like those of hotels and restaurants, become the limiting factor for further growth. And in a wider perspective by looking at the other markets, EVFR and business, we see from a certain volume the contours of market saturation. In developed markets, many people fly in the meantime as much as they need, want of like. Fare decreases from that point onwards will not lead to flying more than these people already do. Aviation therefore has to prepare for low demand growth in developed markets, albeit at large volumes. The US domestic market, the largest single aviation market in the world, shows signs of saturation. Developing markets like China and India will continue to grow with the growth of their middle classes, and the growth of labour migration.

The first limiting factor mentioned above, the cost for other elements of the travel, will likely influence the development of demand for long-haul travel more, simply because most passengers stay longer at the destination in these markets, whereby the other cost of the travel will have a high impact of travel behaviour.

The above teaches us that the generic forecasted demand growth will likely be spread *unequal* in geography and in markets. The general prediction of annual growth between 4% and 5% is not a figure that all players in aviation can use as a planning yardstick as their annual growth may well be lower or higher than the predicted average.

## 18.2 Network development

In Chapter 11 of this textbook we saw that improved fuel efficiency of aircraft leads to aircraft becoming more payload and/or range capabilities. In the meantime the industry operates compact long-haul aircraft like the Boeing 787, Airbus A350 and most notably the Airbus A321-XLR. Such aircraft enable airlines to profitably operate longer thinner routes, increasing the number of non-stop served city pairs. These routes bypass hub airports and will ultimately lead to changes in hub positions of airports. It is fair to assume that the airline business develops towards increasing non-stop connectivity with hub connections becoming secondary.

Networks may well change in another way as well due to the increasing pressure for low fuel burn – read: $CO_2$ emissions – per passenger. Today's long-haul operations involve the need for high amounts of fuel at departure in order to cover the long distance. These amounts of fuel represent a high percentage of the overall aircraft mass at take-off, and the aircraft need fuel to carry all that fuel, hereby eroding the fuel efficiency per passenger on these long routes. It is very well possible that, in order to decrease the amount of fuel per passenger on long-haul routes, airlines may refuel the aircraft half-way the long-haul route. This avoids the aircraft carrying fuel for fuel, and would significantly increase the fuel efficiency per passenger. This could result in airports presently operating as O+D airports to become fuel hubs along the larger long-haul routes.

## 18.3 Capacity constraints

In this textbook, we have at various moments concluded that the growth of aviation in terms of passengers and the number of aircraft deployed grows faster than the growth of airport capacity and airspace capacity. It is therefore inevitable that airport capacity and airspace capacity will be the bottlenecks in aviation in the future ahead of us. Airlines need to consider maximum airport capacities of their home airports for deciding on fleet expansion as we will see situations where the growing fleet of an airline does not fit within the capacity limits of the home airport any longer. Network development, in terms of increase in destinations and frequencies, will foremost become limited by the available airport capacity. The general assumption of airlines opting for very large-capacity aircraft for this reason has materialized very little; the Airbus A380, designed for high capacity operations between capacity-limited airports, is not in high demand. Here we see that aircraft seating capacity has reached a limit in terms of general deployability of such large aircraft and the complexities of disruption management.

When we look at operations of airports today, we can certainly identify room for improvement, resulting in a better deployment of existing airport capacity. Decreasing separation distance or time between aircraft, quicker turnaround processes, streamlining terminal processes, all hot topics in aviation operations today, will certainly lead to significant improvements in the available airport capacity. Streamlining ATC organizations and rationalization of airways by skipping or relocating military corridors will also result in a better deployment of available capacities. These measures will widen the bottlenecks but will not remove them. In due course the newly acquired capacity will be occupied; and as long as aviation is growing at its contemporary rates, capacity constraints become inevitable. It may well result is slowing the pace of growth.

Before that happens, already visible today is the capacity constraints leading to regular and large-scale disruptions. When an industry needs to operate at maximum capacity, the smallest event can be enough for massive disruptions. This results in flight delays or cancellations, increasing the operational cost. The cost for coping with disruptions is already significant in the airline business and will increase in the years ahead of us.

As we have observed in this textbook that aviation is very much a labour driven sector, capacity constraints may arise due to labour market shortages. At the moment of writing we do already see that labour cost in the industry goes up due to labour shortage, and this process will certainly continue. Although educational institutes deliver vast numbers of alumni for the aviation sector, it may not be enough and the numbers are unequally spread over the world. Also, here, the immediate effect to the aviation industry is cost increase: labour cost as the largest cost next to fuel, is likely to go up at most if not all parts of the world.

## 18.4 Fare levels

At various parts of this book we discussed demand, how aviation demand is derived demand with various price elasticities and saturation moments. In general, we can say that the aviation industry is primarily supply driven: airlines buy aircraft on the basis of optimistic assumptions and the ever-growing capacity need to be sold at ever lower fares, made possible by lower *internal* production costs per unit – the supply-driven cocktail for massive growth based on low fares.

But what if fares go up as a result from *external* factors? Governmental bodies today increasingly discuss taxation of fuel or value-added taxes on tickets as a response to the climate challenge. The industry is working towards high-scale production of sustainable fuel, but at significantly higher cost than fossil fuel. Thus, as in the previous paragraph, we observe that internal costs go up as well. This will ultimately result in *fare increases*, and if that happens, the airline business will enter into unknown territories. How will the massive demand, stemming from low fares, develop at increasing fare levels? One needs to be very optimistic for assuming demand not to be influenced downwards by these likely to happen price increases. Many travel by air because it is cheap. How many will continue to fly when it is not cheap anymore? The truth is that nobody knows; it is impossible to analyze because no data are available, and travel behaviour defined by the necessity for travel is changing rapidly.

It is far too simple to assume that the majority of LCC passengers fly for fun and because it is cheap. Many fly based on necessity, such as the massive numbers of labour migrants in uniformed labour markets like the US, the EU and China. Labour migration is probably the largest amplifier for today's massive demand growth. And this is logical, as aviation itself has been and is a massive amplifier for labour migration. Raising fares by adding taxes may well result in a further increase of labour cost, as at the end of the equation the employer pays for the travel of the labour migrant in its labour market.

Travelling for fun is further nuanced by observing that in today's integrated societies, studying at a foreign university of visiting an art exposition in another capital is regarded as the normal way of living in today's society. In short, integrating societies as in the US, China and the EU, as well as in large countries like India, Russia or Brazil, have generated explosive growth of aviation. And again, here as well, it is logical as aviation has been the prime enabler for this way of living.

This all makes it totally uncertain what the effect of fare increase in aviation will be, with the notion that the only certainty is that the effect will not be negligible and will undoubtedly have a negative effect on the projected growth rates.

We have seen in Chapter 4 that demand for aviation is a derived demand, resulting from a demand for transportation. This implies that the mass demand for transportation could be fulfilled by other transport modes. In today's discussions, the fast train is often mentioned as suitable alternative for journeys up to some 1,000 km. In the US as well as in the EU, where the rail alternative is often mentioned, this looks easier than it is. Travel demand is very fragmented between a multitude of metropolises, cities and communities. Many railway systems are saturated with domestic or local demand, and adding all these aviation passengers is simply impossible. Demand transformation from aviation to rail requires vast investments, including fulfilling the vast need for electrical power, and would require uniformity at governmental levels. This latter prerequisite, a fundament for aviation development, is non-existent in rail. In the US, it is corporate and state divided, in the EU railway companies are exclusively national. Given the urge of the climate crisis, where the Paris Agreement stipulates an emission-neutral world by 2050, rail transformation will be limited in possibilities.

This might not be the case with long-haul cargo transports. China is, in collaboration with countries involved, working on developing a high-speed cargo rail between China and Europe, and the effects not only on maritime cargo but also on air cargo may arise sooner than many expect.

Finally, on this, alternatives for aviation are, by definition, surface mode transport systems. For the longer future, surface transport may also appear for longer distances by means of the hyperloop. Presently in its infancy, this energy effective, ultra-high speed transport system may become the basis for an intercontinental subway-type system, bringing masses quickly and cheaply from capital to capital.

## 18.5 Consolidation

Most certainly, the aviation sector will see a continuation of the ever-existing process of consolidation. On the supply side, we see horizontal consolidation at a component level, with the first signs of tier-one OEMs, like Boeing and Airbus, gearing up for vertical consolidation. The future will see fewer less suppliers than there are today.

This will also happen in the airline industry, with the remark that all will depend on the legal developments on airline ownership rules. If these remain unchanged, the airline consolidation will remain and increasingly become a landscape of joint ventures, often with formal ownership division complying with the old rules, and actual control based on financial dominance. If airline ownership rules will become looser or even disappear, we will see multinational airlines appearing. Some joint ventures in the airline business today could be regarded as a line-up to new multinational single entities.

## 18.6 Technology changes

We started this textbook by observing already in the introduction that the development of aviation is foremost defined by technological development, ever increasing engine power against ever increasing fuel efficiency. Will this continue in the decades ahead of us?

Most certainly, it will. The industry is constantly working on further improvement of present technologies, and experimenting with new technologies. We need to observe that improving existing technologies results in marginal improvements in fuel efficiency, and emerging technologies are needed for breakthrough developments. As aviation is a safety-driven industry, application of emerging technologies in aircraft certified for commercial operation will take time. Moreover, we have seen throughout this book that aviation has become very much economy driven. This means that equipment entering the market needs to be deployed by many years in order to recover the investments made. It is fair to say that the world's most efficient aircraft and engines at the moment of writing of this book, say the Boeing 787 and the Boeing 777-9, along with the Airbus A350 and A321-XLR, need some 25 years of production and operation for returning the investment.

And this all implies that full-scale implementation of technological refinements, and even more the application of emerging technologies, will take much time. Given the time window as set in the treaty of Paris, stipulating that the global society needs to be carbon-neutral by 2050, it is extremely speculative to assume that technological developments will enable the aviation sector to grow or even to maintain its present volumes. And this leads to the conclusion that the future of aviation is full with uncertainties for the decades ahead.

# Index

Note: *Italic* page numbers refer to figures.